T0073308

Extremophiles

Koki Horikoshi

Extremophiles

Where It All Began

Koki Horikoshi
Extremobiosphere Research Center
Japan Agency for Marine-Earth Science
    and Technology (JAMSTEC)
Yokohama, Kanagawa, Japan

ISBN 978-4-431-55407-3   ISBN 978-4-431-55408-0   (eBook)
DOI 10.1007/978-4-431-55408-0

Library of Congress Control Number: 2016933121

Springer Tokyo Heidelberg New York Dordrecht London
© Springer Japan 2016
This work is subject to copyright. All rights are reserved by the Publisher, whether the whole or part of
the material is concerned, specifically the rights of translation, reprinting, reuse of illustrations, recitation,
broadcasting, reproduction on microfilms or in any other physical way, and transmission or information
storage and retrieval, electronic adaptation, computer software, or by similar or dissimilar methodology
now known or hereafter developed.
The use of general descriptive names, registered names, trademarks, service marks, etc. in this publication
does not imply, even in the absence of a specific statement, that such names are exempt from the relevant
protective laws and regulations and therefore free for general use.
The publisher, the authors and the editors are safe to assume that the advice and information in this book
are believed to be true and accurate at the date of publication. Neither the publisher nor the authors or the
editors give a warranty, express or implied, with respect to the material contained herein or for any errors
or omissions that may have been made.

Printed on acid-free paper

Springer Japan KK is part of Springer Science+Business Media (www.springer.com)

To my wife, Sachiko

Preface

Imagination is everything. It is the preview of life's coming attractions. Imagination is more important than knowledge. (Albert Einstein)

Recently, I had the chance to see one of the largest exhibitions of Claude Monet's most famous series of paintings in London. As Kenneth Clark explains in his famous book *Civilization*, Monet attempted a kind of colour symbolism to express the changing effects of light. For example, he painted a series of cathedral facades in different lights—pink, blue, and yellow—which seem to me too far from my own experience. The colours of these objects depend on the physical environment, such as sunlight, snow, the time of the day, and the season. Under different conditions, one object may show quite different properties. Who can be sure what is the absolute property? The biological world may have the same uncertainty.

In 1956, I encountered an alkaliphilic bacterium – although not alkaliphilic in the true sense of the word. I was a graduate student in the Department of Agricultural Chemistry, the University of Tokyo, working under the direction of Professor Kinichiro Sakaguchi. Autolysis of *Aspergillus oryzae* was the research theme for my doctoral thesis. The reason Professor Sakaguchi asked me to study the autolysis of *Asp. oryzae* was a practical one. He thought the flavour and taste of Japanese *sake* came from an autolysate of *Asp. oryzae*. Every day, I cultured stock strains of *Asp. oryzae*. After 1 week of culture, all I had to do was taste the cultured fluid.

One day in November, I found one cultivation flask in which the mycelia of *Asp. oryzae* had completely disappeared. The night before, when I had looked at the flasks, the mould was flourishing in all the culture flasks. I still remember vivid pictures of bacteria thriving and moving. No mycelium could be seen under the microscope (Fig. 1 from the first page of my Laboratory notes, 1956).

The microorganism isolated from that flask was *Bacillus circulans*, and strong endo-l,3-β-glucanase activity was detected in the culture fluid. This enzyme lysed *Asp. oryzae*. It was the first time that mould cells had been found to be lysed by bacteria, and these results were published in *Nature* (Fig. 2, Horikoshi and Iida 1958). However, this bacterium showed very poor growth in the absence of mycelia of *Asp. oryzae* and production of endo-l,3-β-glucanase was very low. Therefore,

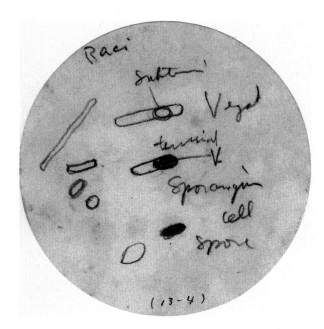

Fig. 1 Contaminated bacteria. This was written in my first laboratory note

Fig. 2 Electron micrographs of cell wall of *Aspergillus oryzae* (**a**) and cell wall after treatment with *Bacillus circulans* enzyme (**b**)

purification of endo-1,3-β-glucanase could be done only in culture fluid in the presence of mycelia of *Asp. oryzae*. I did not realize at the time that the culture fluid had an alkaline pH value. A few years later, I attempted production of endo-1,3-β-glucanase in conventional media. I tested many culture media containing various nutrients. The addition of 0.5 % sodium bicarbonate to conventional nutrient culture broth gave good growth and production of the enzyme. Autolysis of *Asp. oryzae* changed the culture medium from weakly acidic to alkaline pH. I discovered that such a change in pH value accelerated bacterial growth.

In 1968, on a visit to Florence, Italy, I greatly admired the Renaissance buildings, so very different from traditional Japanese architecture. At the time of the great flourishing of the Renaissance in Italy – the fifteenth and sixteenth centuries – half a world away in Japan it was a time of civil wars and turmoil. Nevertheless, it was the age that saw the development of the tea ceremony, Noh theatre, and the art of gardening, as well as the emergence of ink painting (*sansuiga*). And yet at the time, it would have been impossible for a Japanese to have imagined the extraordinary achievements taking place on the other side of the globe. Here were two remarkable cultures, utterly different, each springing from its own particular circumstances of politics, civilisation, climate, tradition, etc. – in a word, its own medium (Fig. 3).

Of course, to a microbiologist, the words "culture" and "medium" have more than one meaning, and perhaps it was this that prompted a new idea in my mind. Could this kind of extreme diversity be reflected in the microbial world? Was it possible, as a voice seemed to whisper in my ear, that there might exist an entire undiscovered domain of unknown organisms in different, unexplored cultures?

Fig. 3 Duomo in Firenze

Memories of experiments on *B. circulans* done years before flashed into my mind. Could such a domain of microorganisms exist at alkaline pH? The acidic environment was being studied, probably because most food is acidic. However, very little work had been done in the alkaline region. Upon my return to Japan I prepared an alkaline medium containing 1 % sodium carbonate, put small amounts of soil collected from various areas of the Institute of Physical and Chemical Research (RIKEN), Wako, Japan, into 30 test tubes, and incubated them overnight at 37 °C. To my surprise, microorganisms flourished in all the test tubes. Figure 4 is the first laboratory note of alkaliphiles. I isolated a great number of alkaliphilic microorganisms and purified many alkaline enzymes. The first paper concerning an alkaline protease was published in 1971 (Horikoshi 1971a).

Then, in 1972, I was talking with my father-in-law, Shigeo Hamada, about alkaliphilic microorganisms. He had been in London almost a century ago as a businessman and was curious about everything. He showed interest in alkaliphiles. As I was speaking, he said, "Koki, wait a minute, I have an interesting present for you." He brought out a page of an old newspaper, *Nikkei Shimbun*, dated June 11, 1958. A short column with an illustration of an electron micrograph was like a punch to my head. I had not known this! The article said:

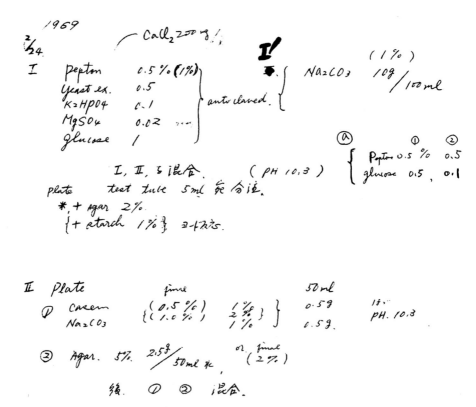

Fig. 4 The first laboratory note of alkaliphiles, dated 24 February 1969

In Japan, since ancient times, indigo has been naturally reduced in the presence of sodium carbonate. Indigo from indigo leaves can be reduced by bacteria that grow under high alkaline conditions. Indigo reduction was controlled only by the skill of the craftsman. Takahara and his colleagues isolated the indigo reducing bacterium from a ball of indigo.

I then carefully checked scientific papers from *Chemical Abstracts* in the library of RIKEN. Only 16 scientific papers on alkaline-loving bacteria were discovered. These bacteria remained little more than interesting biological curiosities. No industrial application was attempted at all (Horikoshi and Akiba 1982). I named these microorganisms that grow well in alkaline environments "alkaliphiles" and conducted systematic microbial physiological studies on them. It was very surprising that these microorganisms, which are completely different from any previously reported, were widely distributed throughout the globe (even at the deepest point of the Mariana Trench in the Pacific Ocean), producing heretofore unknown substances. Here was a new alkaline world that was utterly different from the neutral world.

Over the past four decades my coworkers and I have focused on the enzymology, physiology, ecology, taxonomy, molecular biology, and genetics of alkaliphilic microorganisms to establish a new microbiology of alkaliphilic microorganisms. A big question arises, "Why do alkaliphiles require alkaline environments?" The cell surface of alkaliphiles can maintain intracellular pH values about 7–8 in alkaline environments of pH 10–13. How the pH homeostasis is maintained is one of the most fascinating aspects of alkaliphiles. To understand this simple but difficult question, we carried out several basic experiments to establish gene recombination systems (Horikoshi and Grant 1998). Finally, after almost 2 years, the whole genome sequence of alkaliphilic *Bacillus halodurans* C-125 was completed. This was the second whole-genome sequence of spore-forming bacteria thus far reported. This sequence work revealed interesting results. Many genes were horizontally transferred from different genera and different species as well. However, we still have not found the crucial gene(s) responsible for alkaliphily in the true meaning (Horikoshi 2006).

Industrial applications of these microorganisms have also been investigated extensively, and some enzymes, such as alkaline proteases, alkaline amylases, alkaline cellulases, and alkaline xylanases, have been put to use on an industrial scale. Subsequently, many microbiologists have published numerous papers on alkaliphilic microorganisms in various fields. At the beginning of our studies, very few papers were presented, but now thousands of scientific papers and patents have been published. It is not clear which field our study of alkaliphiles will focus on next, but the author is convinced that alkaliphiles will provide much important information (Horikoshi 2011).

Time is the best appreciator of scientific work, and we know that an industrial discovery rarely bears all its fruits in the hands of its first inventor. (Louis Pasteur)

Other Extreme Environments

Not too many years ago, almost all biologists believed that life could survive only within a very narrow range of temperature, pressure, acidity, alkalinity, salinity, and so on. Nature, however, contains many extreme environments, such as hot springs, saline lakes, deserts, alkaline or acidic lakes, and the deep sea. All these environments would seem to be too harsh for life to survive.

However, in recent times many organisms have been found in such extreme environments. The idea of extreme environments is relative, not absolute (Brock 1997; Brock and Freeze 1969). Clearly we have been too anthropocentric in our way of thinking. We should therefore extend our consideration to other environments having multiple extreme conditions for life. Such organisms metabolise inorganic sulfur or iron as their energy source. Some of these isolated from deep- or sub-deep-seabed sediment have an entirely different metabolic pathway from conventional life. Figure 5 shows the deepest point on the Earth, and Fig. 6 shows shrimps at a depth greater than 10,000 m.

It is distinctly possible that very ancient life forms may be in hibernation in the deep ocean, the world's largest refrigerator.

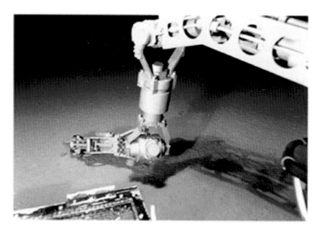

Fig. 5 Sampling at a depth of 10,898 m

Fig. 6 Shrimps below 10,000 m

Distribution of Polyextremophiles

On the wider scale, extreme environments on Earth have arisen and continue to arise as a consequence of plate tectonic activity, the dynamic nature of the cryosphere, and the formation of endorheic basins. Plate boundaries occur wherever two tectonic plates collide and result in the formation of mid-ocean ridges, mountains, deep-ocean trenches, volcanoes, and other geothermal phenomena such as marine hydrothermal vent systems.

Hydrothermal vent systems are found in abundance worldwide and are presumed to have existed as soon as liquid water accumulated on Earth. Black smoker and carbonate chimney vents are different environments: black smokers arise at diverging plate boundaries above magma chambers, are highly acidic (pH 1–3), and are very hot (up to 405 °C), with vent fluids rich in Fe and Mn, and CO_2, H_2S, H_2, and CH_4. In contrast, carbonate chimneys found off-axis (away from diverging boundaries), are highly alkaline (pH 9–12), moderately hot (up to 90 °C), and rich in H_2, CH_4 and low molecular weight hydrocarbons.

A high proportion of the Earth's surface contains water in solid form (sea ice, ice caps and sheets, glaciers, snowfields, permafrost), the longevity of which may be thousands of years or even a few million years. Cryosphere–climate dynamics are complex and influence precipitation, hydrology, and ocean circulation. Deserts develop in regions where precipitation is very low (or zero) and unpredictable. Highly saline lakes and pans often develop in these circumstances (Horikoshi 2011).

The average depth of the world's oceans is about 3,800 m; high pressure generates yet another extreme environment. Oligotrophic environments are defined as those presenting very low nutrient concentrations; they include oceans depleted in iron, nitrate or phosphate, tropical laterite soils, and white sands. Finally, a range of environments are deemed to be extreme by virtue of chemically or physically caused toxicity (e.g., soils high in arsenic, lakes exposed to high incident radiation).

New extreme ecosystems continue to be discovered and investigated, including the deep biosphere that exists at great depths in sub-seafloor sediments, in subterranean rock formations, and in the carbonate chimney vent system. Extreme environments almost invariably are affected by two or more extreme conditions. Is our current knowledge of extremophile diversity comprehensive? It is highly possible that acidopsychrophiles, acidohalophiles, and thermohalophiles exist, and these should not be neglected by microbiologists.

It may appear overly hasty to introduce evolution and discussion of the origin of life in the Preface. Assuming a thermophilic beginning, acidophily probably arose at an early stage, whereas alkaliphily evolved only after precipitation of certain minerals and a sufficient buffer concentration of CO_2 were established in the atmosphere. Moreover, halophily could have developed only after an arid climate was imposed on land and psychrophily only after a major temperature fall. This is speculation, of course, but for me – and, I hope, for students entering the world of microbiology and extremophiles – the enduring fascination of this field of study is the huge contribution that these microscopic organisms can make to our understanding of the grand story of life on Earth.

The question "what is life?" is precisely the question "what is evolution?" (Carl R. Woese)

Tokyo, Japan   Koki Horikoshi

Acknowledgements

The author expresses sincere gratitude to his wife, Sachiko, for her invaluable help during the past 50 years, and is also grateful to Mr. Peter Ingham for his assistance in finalising the English manuscript.

Tokyo, Japan
August 2014

Contents

Chapter 1
Prologue

"Studying the Past to learn New Things," Analects of Confucius, 551 to 479 BC, China. A Chinese social philosopher, whose teachings deeply influenced East Asian life and thought.

There are many books on the life sciences, but unfortunately younger scientists seldom have the chance to study the history of biology before the 1950s. For example, they have few opportunities to learn about the historic indigo dying process or sake brewing, and yet there are a number of traditional fermentation techniques, such as mixed fermentation and pasteurisation of sake.

The Japanese way of thinking is compared with the reductionism derived from Western culture in this book. For younger readers, the applications of biology are also listed.

Are the Japanese capable of originality in the field of biology? Yes, we have creative power, although our concept is different. Japan was closed to foreigners for a long period, but we had many private elementary schools during the Edo period (eighteenth to nineteenth centuries). The pupils had the chance to learn academic subjects. Mathematics was a kind of hobby, and if someone found a new solution, he would offer a votive wooden plaque at a shrine. However, I am sorry to say, Japanese students of modern biology lack originality.

One of the oldest and best established biotechnologies in Japan is sake brewing. As is well known, the Japanese climate is naturally warm and humid. Many types of mould, yeast, and bacteria flourish all over Japan, but how did our ancestors harness these to establish the sake-brewing process? They did not isolate any microbes. Instead, the secret of sake brewing was discovered through the Japanese way of thinking. This sensibility still resides in the Japanese and is present in me as a biologist.

A story published in the 1960s tells of the serendipitous discovery of R-factor.

In Gunma Prefecture, farmers used to meet for afternoon tea. At these occasions it was customary to serve homemade pickles rather than cakes. It transpires that these pickles were infected with antibiotic-resistant *Escherichia coli*. This resistance, it turned out, could be transmitted to a dysentery bacillus widespread at the time.

© Springer Japan 2016
K. Horikoshi, *Extremophiles*, DOI 10.1007/978-4-431-55408-0_1

Professor Mitsuhashi found that many patients carried antibiotic-resistant *E. coli* transferred by R-factor (episome) to the antibiotic-resistant dysentery bacillus. He told me: "If we had not had so many dysentery patients and tea parties with pickles contaminated with drug-resistant bacteria, I would not have been able to discover this small DNA fragment."

In turn, I was inspired by the Renaissance culture of Florence, so different from the Japanese Muromachi period occurring at the same time (fourteenth to sixteenth centuries). It was this inspiration that led me to the discovery of the new biology of "alkaliphiles."

My research developed further to deep-sea and subsurface life. However, at that time neither a suitable journal nor a scientific society existed for scientists studying extremophiles. Consequently, I founded the journal *Extremophiles* and the International Society for Extremophiles.

There are no borders in science, but there are kinds of borders in the line of thinking derived from different religions and different languages. Even though we may reach the same result, it might well be by a different way of thinking. The diversity of cultures and ways of thinking can promote the development of science. Therefore, international collaborations and exchanges of information are absolutely necessary for all investigators.

Science is the one common language of all human beings. We have just started to communicate with the universe by using science. Science is just a blank sheet of paper. If Michael Faraday were to write on this paper, the paper would become the *Chemical History of a Candle*. If Charles Darwin were to write on the paper, the paper would become *On the Origin of Species*. I am convinced that we will have the opportunity to understand what life is through science.

So, I urge younger scientists: You can learn much from different cultures in different countries.

> Those who explore an unknown world are travellers without a map; the map is the result of exploration.
>
> The traveller, Hideki Yukawa, Nobel Prize in Physics (1949)

Chapter 2
Early Microbiology

Sake Brewing Was the First Biotechnology in Japan

Brewing Japanese sake is one of the most important crafts handed down from ancient times. Fermentation was the largest chemical enterprise and made use of excellent biological technology. It is no exaggeration to say that sake making was the result of the combination of science and technology even in those days.

For a long time, sake production was a government monopoly, but in the tenth century temples and shrines began to brew sake, and they became the main centres of production for the next 500 years. *The Diary of Goshu* (1489) and *The Tamon-in Diary*, written by the abbots of Tamon-in (temple) from 1478 to 1618, record many details of the brewing technology used in the temple. The diary shows that the process now known as pasteurisation and the process of adding ingredients to the main fermentation mash in three stages were established practices by that time.

These books describe three fundamental technologies, which are still generally accepted today (Figs. 2.1 and 2.2).

1. How to make *koji* (starter culture).
2. How to isolate pure yeast (*syubo*, *Saccharomyces cerevisiae*) using lactic acid bacteria.
3. How to pasteurise sake at 50–60 °C to protect against decay.

This protocol was established about 300 years ago, and essentially these three technologies have changed little up to the present day.

Ancient Sake Brewing

Japan enjoys a relatively high temperature and a humid climate. Rice left at room temperature for a week will fall prey to many types of mould, yeast, and bacteria.

© Springer Japan 2016
K. Horikoshi, *Extremophiles*, DOI 10.1007/978-4-431-55408-0_2

Fig. 2.1 Sake brewing (washing rice) about 200 years ago. A woodcut written in older Japanese

Fig. 2.2 Sake brewing (koji making) about 200 years ago. A woodcut written in older Japanese

How were our ancestors able to isolate the *koji* mould *Aspergillus oryzae* and the yeast *Saccharomyces cerevisiae*? They did not, of course, have microbial knowledge, but isolated the *koji* mould and *syubo* yeast by making use of the struggle for existence in nature. Even in the context of modern microbiology, this technique is unique.

1. Pure seed *koji* (*Aspergillus* (*Asp.*) *oryzae*)

   Starch in rice should be hydrolysed to glucose by *Asp. oryzae*. Our ancestors made seed, the conidiospores of *Asp. oryzae*, for sake brewing by culturing *Asp. oryzae* on steamed rice mixed with wood ash at room temperature. After 1–2 weeks, almost pure conidiospores could be collected as seed *koji*. The addition of ash gave the solid culture a high concentration of minerals and elevated pH values. In these conditions many species of mould and bacteria were unable to grow or grew very slowly, with the exception of *Asp. oryzae*. Seed *koji*, of course, differed from place to place, and this difference produced different sakes. People in Kyoto, for example, liked to drink sake made in Kyoto, which used the ash of trees from the southern part of Kyoto.

2. The struggle for existence between lactic acid bacteria and yeast

   One of the most important microorganisms in sake making is the yeast *Saccharomyces cerevisiae*.

   Sake used to be made in winter (at temperatures lower than 10 °C). *Koji* mould hydrolysed rice starch and produced glucose. Many species of microbes tended to infect the steamed rice containing *koji* mould. However, *Lactobacillus* grew rapidly at low temperature, and the activity of lactic acid lowered the culture pH. Almost all microbes were killed except the yeast *Saccharomyces cerevisiae*. Then, the culture fluid (*moromi*, main mash) was warmed to 20 °C and stirred vigorously. The lactobacillus was killed by the high temperature and oxygen. Consequently, only *Saccharomyces cerevisiae* could grow and thus yield sake. Here, we had a beautiful example of how the struggle for existence among microbes could be harnessed to productive ends.

3. Pasteurisation at 55 °C

   Another surprising technology was pasteurisation at relatively low temperatures (*hiire*). Of course, the ancients did not have thermometers and measured temperature by touch. *Hiire* started at the end of the sixteenth century, about 300 years before Louis Pasteur discovered pasteurization. Although the Japanese did not know about microorganisms, they understood that high temperatures killed microbe infections. We had to wait for Pasteur's work to explain the science of *hiire*.

Our ancestors isolated *koji* mould and yeasts by controlling their environments and thus developed the biotechnology necessary for brewing sake. This traditional heritage remains the foundation of modern biotechnology even today.

Indigo Fermentation

In Japan, the reduction of indigo for dyeing is a traditional use of alkaliphilic bacteria dating back to ancient times. Indigo from indigo leaves was reduced by alkaliphilic bacteria in an alkaline environment containing sodium carbonate, calcium hydroxide, and potassium hydroxide. After the discovery of synthetic indigo, indigo was reduced to indigo white by the addition of reducing agents such as hydrosulfite. All dye factories have used this chemical reduction process.

Recently, traditional indigo dyeing has been revived as a folk craft. There are many traditional indigo dyeing processes in Japan, one such is as follows:

1. Wood ash (17 kg), slaked lime (4.4 kg), and 160 l hot water are mixed, stirred for 3 h, and allowed to stand for 48 h at room temperature. About 140 l supernatant fluid (ash extract) is obtained.
2. About 3 kg indigo balls (*tsukumo* of *aidama*, fermented indigo plant *Polygonum tinctorium*) is mixed with 25 l ash extract and ground with a pestle and mortar. Then, approximately 40 l ash extract is added to the mixture. The pH of the mixture is about 11.2–11.7.
3. The mixture is kept at 20–30 °C for 4 days and 20–40 l of ash extract is added at 24-h intervals to maintain high pH values.
4. As nutrient for the alkaliphilic bacteria, 210 g wheat gluten is added. The pH of the mixture is then adjusted by the addition of potassium hydroxide.
5. After 8 days, fermentation starts and foam is observed. About 400 ml sake (rice wine) is added as a booster.
6. During the fermentation process, the pH value is kept between 10.0 and 11.0 by the addition of potassium hydroxide.
7. After 2 weeks, the reduction of indigo is complete and the foam of reoxidized indigo is observed. Figure 2.3 shows this traditional method (about 150 years ago) of indigo reduction.

Later on, indigo-reducing *Bacillus* sp. no. S-8 was isolated from an indigo ball and used to develop an improved indigo reduction process on an experimental scale. By adding a seed culture of *Bacillus* sp. no. S-8 to the reduction mixture, the processing time was decreased by 75 %. Reduction was easier, and the product was considered better than that obtained by the traditional fermentation method.

Furthermore, many different microbes have been derived from soils in different ateliers. Different microbes have different reduction pathways of indigo, and many shades were dyed on fabrics. These experiments were carried out before the revival of traditional indigo dyeing, and few microbiologists noted these results.

Cultures of Moulds and Malt

Jyoukichi Takamine was born in Takaoka, Toyama Prefecture, in November 1854. His father was a doctor and his mother a member of a family of *sake* brewers. He was educated in Osaka, Kyoto, and Tokyo, graduating from the Tokyo Imperial

Fig. 2.3 Traditional indigo reduction in Hanyu City, Saitama Prefecture, where I was born. (Woodcut made about 150 years ago)

University in 1879. He did postgraduate work at the University of Glasgow and Anderson College in Scotland. He returned to Japan in 1883 and joined the chemistry division at the Department of Agriculture and Commerce.

Takamine continued to work for the department until 1887. He then founded the Tokyo Artificial Fertilizer Company, where he later isolated the enzyme takadiastase, an enzyme that catalyses the breakdown of starch in a fungus (*Aspergillus oryzae*).

He later emigrated to the United States and established his own research laboratory in New York City, and licensed the exclusive production rights for takadiastase to one of the largest U.S. pharmaceutical companies, Parke Davis. He became a millionaire in a relatively short time and by the early twentieth century was estimated to be worth $30 million.

In 1901 he isolated and purified the hormone adrenaline (the first effective bronchodilator for asthma) from animal glands, becoming the first to accomplish such a feat for a glandular hormone.

There are several differences between sake brewing and the making of other alcoholic drinks. In Japan the starch of rice is hydrolysed by mould enzymes, whereas in Western countries malt enzymes are used to convert starch to glucose.

In 1894, Takamine succeeded in producing takadiastase from the mould *Aspergillus oryzae* that was much stronger and less expensive than malt diastase. He hoped to use his takadiastase not only for digestive but also for industrial applications. However, people in the USA and Europe refused to consider any applications using mould diastase because *koji* mould was not popular.

About 40 years ago in California I had an interesting experience. My mother had kindly sent us some rice cakes by sea mail. Upon arrival we found that the rice cakes that had originally been white had turned green. The green powder was a conidiospore of *koji* mould. *Koji* mould grew on steamed rice, bread, and many foods in our house. However, after 2 or 3 weeks the amount of *koji* mould decreased and after 1 month had disappeared completely. Japanese *koji* mould could not survive in the California environment.

As for Takamine's takadiastase, he planned to use it as a less expensive means to hydrolyse the starch of corn and to make bourbon whiskey. He met ferocious opposition from malt makers and workers because fungus enzymes were unfamiliar to almost all of them. They associated fungi with dirt. Finally, his plant in Illinois burnt down and with it his plans to make bourbon.

Takadiastase, however, was imported in Japan, and several enzymes were isolated. A number of biotechnology companies were established using technology learnt from the production of takadiastase

As an interesting footnote, the cherry trees in the West Potomac Park in Washington, D.C., were donated by the mayor of Tokyo, Yukio Ozaki, and Dr. Takamine in 1912.

Cheese and Soy Sauce

The first step in cheese making is the curdling of milk proteins, producing a solid mass from which much of the water is drained away. In traditional cheese making, however, the curdling has long been triggered by rennin (now known as chymosin), which is a protein-degrading enzyme usually extracted from the stomach of calves.

Increasingly, however, and particularly in the United States, cheese has been made using enzymes manufactured by the mould *Mucor miehei*, isolated by Professor Kei Arima of the University of Tokyo in 1960 (Fig. 2.4), which has an action similar to that of chymosin. Although suitable for producing quick-ripening cheese – and vegetarian cheese, which has become popular – the enzymes from *Mucor miehei* are not entirely satisfactory because their activity reduces relatively quickly.

More recently, genetic engineers have transferred the gene responsible for chymosin production into bacteria, yeast, or mould to manufacture chymosin. For more traditional cheese fanciers, however, microbial sources of chymosin may never entirely replace that obtained from a calf's stomach, which contains traces of other enzymes that have significant effects on flavour during the ripening process.

From both a microbiological and gastronomic perspective, cottage cheeses and cream cheeses are the simplest of all. They are made in a single process by adding bacteria such as *Leuconostoc* to pasteurised milk. Lactic acid formed by the bacterial enzymes precipitates the curd.

Most cheeses have to be ripened by bacteria and fungi. Although initially almost all natural cheeses look alike, differences in these microbes and the changes effected

Fig. 2.4 *From the left:* Professor Tadahiko Ando, Professor Kei Arima, Professor Gakuzo Tamura, Koki, and Professor Kazuo Izaki

by their enzymes create a wide diversity of cheeses. *Penicillium roqueforti* is responsible for ripening one of the most venerable of hard curd cheeses, roquefort, made from the milk of ewes. The same fungus is also the principal ripener of stilton and gorgonzola cheeses and the source of their characteristic blue veins.

In cheddar, the same bacteria that (together with rennin) produce the curd also ripen the cheese. As they die, their cells release enzymes, which, working on the milk fat and proteins, form the many different compounds that give cheddar its characteristic flavour. The nutritive value of the cheese also rises greatly, as the bacteria synthesise vitamins, particularly those of the vitamin B complex. In the early weeks of ripening, the number of bacteria reaches hundreds of millions per gram.

Soft and semi-soft cheeses such as camembert and limburger owe their consistency and flavour to microbes whose enzymes soften the curd during ripening. A cosmopolitan population of bacteria, moulds, and yeasts lives on the surface of the cheese in a slime that contains as many as ten billion microbes per gram. Their enzymes diffuse into the cheese, softening it and creating its characteristic taste and aroma. *Penicillium camemberti* is the chief microbe in the ripening of camembert. Many others are not merely tolerated but welcome: the outer skin of camembert contains a massive number and assortment of microbes. Traditionally, cheese makers inoculate new batches with a surface smear from an older cheese, but do not seek to suppress other organisms. Increasingly, however, carefully nurtured "starter cultures" are used.

Back in the 1950s, a group of scientists succeeded in making something that, in appearance and chemical composition, was identical to cheddar cheese. They started with sterile milk, devoid of all microbes, and added a chemical (gluconic acid lactone) to precipitate the curd. But their work was soon justly forgotten, consigned to the annals of misplaced science. The cheddar substitute they produced had no cheese flavour whatsoever.

Miso and Soy Sauce

In 1958, while I was working in a microbiology lab in Purdue University, Lafayette, Indiana, I tried to make *miso* paste, as it was difficult to buy the paste for making *miso* soup except in Chicago, Illinois, about 150 km from the university. *Miso* paste is made very simply in Japan, using just boiled soya beans, rice *koji* (rice and *Aspergillus oryzae*), and salt. The mixture is kept for 1 or 2 weeks at room temperature. In Indiana, however, several grams of *miso* paste made in Japan was added to boiled soya beans as a starter. After a few weeks ripening, a splendid *miso* paste was then produced. In this way I came to understand that the Japanese miso-starter contained various microbes besides *Aspergillus oryzae*.

Another important seasoning is soy sauce. The raw materials are almost the same as those of *miso* paste: boiled soya, roasted rice or wheat *koji*, salt, and water. After 6–8 months fermentation, these raw materials are hydrolysed and converted by fungi, yeasts, and lactic bacteria in soy-making plants and change into soy sauce with its characteristic colour, taste, and flavour. If the microbes in the plants were different, would the taste of the soy sauce change? I asked my friend Dr. M. Nagasawa, who was the first plant manager of Kikkoman Co. in Wisconsin in 1973. He gave me a very revealing answer. The main Kikkoman plant is situated in Chiba Prefecture, near Tokyo. When the company established a subsidiary plant in the Kyoto area, it was unable to brew the same soy sauce as was made in Chiba. Eventually, they were able solve this problem by the addition of starter soy sauce from Chiba. Such a starter system meant that they had no trouble producing soy sauce in the USA, he said. Soy sauce is now an absolutely essential ingredient in Japanese cuisine all over the world.

The Korean War and Industrial Fermentation of Amino Acids

In 1908, Kikunae Ikeda, a chemistry professor at the University of Tokyo, isolated glutamic acid as a new taste substance from the seaweed *Laminaria japonica*, more commonly called *kombu*, by aqueous extraction and crystallisation, and called this taste "*umami*." He noticed that the Japanese broth of *katsuobushi* and *kombu* had a peculiar taste that had not previously been scientifically described and which differed from sweet, salty, sour, and bitter. To verify that ionised glutamate was responsible for the *umami* taste, Ikeda studied the taste properties of many glutamate salts such as calcium, potassium, ammonium, and magnesium glutamate. All the salts elicited *umami* in addition to a certain metallic taste because of the presence of other minerals. Among these salts, sodium glutamate was the most soluble and palatable, and crystallised easily. Ikeda named this product monosodium glutamate (MSG) and submitted a patent to produce MSG. The Suzuki brothers started the first commercial production of MSG in 1909 as *Aji-no-moto*, meaning "essence of taste."

Glutamic Acid Production by Fermentation

In 1957, Shikuro Kinoshita and Shigezou Udaka, microbiologists in Kyowa Fermentation Industry, pioneered large-scale industrial fermentation. MSG (monosodium glutamate) was already on the market before the Second World War but only in Japan. In June 1950, war broke out on the Korean peninsula. The U.S. Army had to produce large amounts of canned food containing MSG as seasoning, which was the first export of MSG to the world. At that time, Japanese companies produced it only from the hydrolysate of soya bean protein. Because of a shortage of raw materials, companies were looking for alternative production processes. Udaka, who had been a student of the microbiologist Professor Kin-ichiro Sakaguchi at the University of Tokyo, believed in the diversity and power of microorganisms. Kinoshita and Udaka isolated huge numbers of bacteria from Japanese soil samples (Kinoshita et al. 1957). They cultured them in fluid containing starch and tried to convert the starch to MSG (Fig. 2.5). In the process, they were able to isolate one bacterium that produced large amounts of glutamic acid in its culture broth. Glutamic acid was an essential amino acid for building up the cells. Why did it secrete such an important substance? He found that trace amounts of biotin (a B-group vitamin) and large amounts of ammonium were required for the secretion of MSG. Genetic mutation and improved culture conditions (pH values and aeration, etc.) could produce high yields of MSG in industrial-scale plants.

Udaka thought that, as the microorganisms incorporated nutrients that were metabolised by enzymes on a metabolic pathway and yielded energy, if at some point the pathway narrowed, metabolic compounds would overflow. Consequently, in a culture fluid containing biotin, MSG was secreted to the outside of cells.

At almost the same time another microbiologist, Dr Akira Kuninaka, realised that the ribonucleotide guanosine monophosphate present in *shiitake* mushrooms also conferred the *umami* taste. One of Kuninaka's most important discoveries was the synergistic effect between ribonucleotides and glutamate. When foods rich in glutamate are combined with ingredients that have ribonucleotides, the resulting taste intensity is higher than the sum of both ingredients.

These findings elevated the Japanese fermentation industries to the top level, and this approach was also applied in other amino acid production plants.

J. Gen. Appl. Microbiol.
Vol. 3, No. 3, 1957

STUDIES ON THE AMINO ACID
FERMENTATION

PART I.   PRODUCTION OF L-GLUTAMIC ACID
BY VARIOUS MICROORGANISMS*

SHUKUO KINOSHITA, SHIGEZO UDAKA, MASAKAZU SHIMONŌ

Tokyo Research Laboratory, Kyowa Fermentation Industry Company

Received for publication June 27, 1957

INTRODUCTION

The advantage of microbial methods for the preparation of amino acids is that the product is expected to be purely optically active. Our primary object was concerned with the fermentative production of L-glutamate, which has a big commercial demand as a flavoring agent and has been supplied wholly from plant resources.

It has been generally recognized that glutamate is one of the primary products of nitrogen metabolism in the living cell and the glutamic dehydrogenase system represents an important link between the metabolism of amino acids and carbohydrates[1]. It is also known that glutamate formed in such a way is apt to be transformed rapidly to various other amino acids and proteinaceous materials. For this reason, the production of glutamate directly from carbohydrate and ammonia sources has not been considered seriously before. Since the fermentative production of α-ketoglutarate in a high yield has become possible, several preparative methods of L-glutamate from α-ketoglutarate were reported in academic communications as well as in patent literatures[2,3,4].

Amino acids including glutamate, which were formed in synthetic media containing glucose and inorganic nitrogen sources by various microorganisms, have been frequently detected in a minute amount inside the cell or in the surrounding medium[5,6]. However, the accumulation of glutamate in culture medium in a large amount directly from carbohydrate and ammonia source has never been described, until very recent reports by Asai *et al.*[7] appeared. Asai *et al.* reported the fermentative production of glutamate by various microorganisms. Independently from these workers, an extensive work on this subject was undertaken by the authors and it was found that L-glutamate as well as certain other amino acids could be produced in a good yield directly from carbohydrate and ammonia sources by various selected

* This work was presented at the annual meeting of the Agricultural Chemical Society of Japan held on April 10, 1957.

Fig. 2.5 The first scientific paper on the production of monosodium glutamate by microorganisms

Afternoon Tea and Genetically Modified Bacteria: Discovery of Episome

After the Second World War, dysentery was widespread in Japan because of insanitary conditions. Fortunately, the discovery of antibiotics decreased the death rate dramatically. However, in the 1950s it was reported that the dysentery bacillus which had been responsive to antibiotics could easily become antibiotic resistant.

Although doctors strongly recommended patients to continue taking antibiotics until they had recovered completely, once the diarrhoea eased many patients stopped taking them because they were very expensive.

In 1959, Doctors Asaichiro Akiba and Sadao Kimura of the University of Tokyo discovered something very strange. They isolated an antibiotic-resistant *Escherichia coli* from people who were neither infected with dysentery nor had taken any antibiotics. The antibiotic-resistant *E. coli* made the antibiotic-sensitive dysentery bacillus antibiotic resistant simply by the mixture of the two. Mixing resistant dysentery bacillus with sensitive *E. coli* made the *E. coli* resistant. They found that simply mixing in test tubes could transfer antibiotic resistance.

Discovery of the R Factor

In the 1950s, farmers in Gunma Prefecture were in the habit of visiting each other for afternoon tea. They usually served homemade pickles (not cakes). These pickles were infected with antibiotic-resistant *E. coli*. If anyone was suffering from dysentery, antibiotic resistance would be transferred to the dysentery bacillus from the *E. coli*. Consequently, patients suffered from antibiotic-resistant dysentery.

A report by Professor Susumu Mitsuhashi, of Gunma University, described how transmissible drug resistance factor R (Fig. 2.6), which confers resistance to tetracycline, chloramphenicol, streptomycin, and sulfonamide, was previously found to be transduced in the system of the *Salmonella* E group with episome. The R factor of R^+ transductants was nontransmissible by cell-to-cell contact, and it was not eliminated by treatment with acridine dye. When R^+ transductants were infected with F factor, the nontransmissible R factor acquired transferability by conjugation. The R^+ conjugants, to which only the R factor was separately transmitted by conjugation from the (F^+R^+) donor, were still unable to transfer their R factor by conjugation. However, the $(FR)^+$ conjugants, to which both F and R factor were transmitted simultaneously by conjugation, were also capable of transferring their F and R factors by conjugation.

From this study, it was concluded that the recombinant (FR) factor was formed as a result of an interaction between F and R factors present in a host bacterium, and that one of the mechanisms of acquisition of transferability was accounted for by the formation of recombinant (FR) factor. The recombinant (FR) factor was transferable by conjugation, and it conferred both the drug resistance and F^+ characters

DRUG RESISTANCE OF ENTERIC BACTERIA

II. Transduction of Transmissible Drug-Resistance (R) Factors with Phage Epsilon[1]

KENJI HARADA, MITSUO KAMEDA, MITSUE SUZUKI, and SUSUMU MITSUHASHI

Department of Microbiology, School of Medicine, Gunma University, Maebashi, Japan

Received for publication 3 September 1963

ABSTRACT

HARADA, KENJI (Gunma University, Maebashi, Japan), MITSUO KAMEDA, MITSUE SUZUKI, AND SUSUMU MITSUHASHI. Drug resistance of enteric bacteria. II. Transduction of transmissible drug-resistance (R) factors with phage epsilon. J. Bacteriol. 86:1332–1338. 1963.—Transmissible drug-resistance (R) factors, which transfer resistance to tetracycline (TC), chloramphenicol, streptomycin, and sulfonamide by cell-to-cell contact, were found to be transduced in the system of *Salmonella* E group with phage epsilon (ϵ_{15} and ϵ_{34}). The R+ transductants of *S. newington* (S-84) and *S. chittagong* (S-224) were all found to be unable to transfer their R factors by conjugation, and their R factors were not eliminated by treatment with acridine dyes so far as tested. The R factors containing TC resistance were consistently segregated when transduced. At low multiplicities of infection, the R+ transductants with ϵ_{15} were all nonlysogenic and unable to produce normal ϵ_{15} phage particles; among the R+ transductants with ϵ_{34}, 34% were lysogenic and 66% were sensitive to ϵ_{34}.

It was found that multiple drug resistance was transferred in vitro from resistant *Escherichia coli* to shigellae, and also from resistant shigellae to *E. coli* (Ochiai et al., 1959; Akiba et al., 1960). We confirmed this finding and demonstrated that this transmission is not mediated by transduction, transformation, or a filtrable agent, but by cell-to-cell contact (Mitsuhashi, Harada, and Hashimoto, 1960; Harada et al., 1961). This was also confirmed by blender treatment (Watanabe and Fukasawa, 1960a, b). It was proposed to designate as "R" (resistance) this

[1] The outline of this paper was published in Med. Biol. (Tokyo) 62:13–16, 1962 (in Japanese).

transmissible drug-resistance factor (Mitsuhashi, 1960).

The transmissible drug-resistance factor can also be transferred by transduction in *E. coli* K-12 with phage P1kc (Nakaya, Nakamura, and Murata, 1960; Watanabe and Fukasawa, 1961b). In a previous paper (Mitsuhashi, Harada, and Hashimoto, 1961a; Mitsuhashi et al., 1962), it was reported that recombinant R factors were formed when two kinds of R factor were brought together in a host bacterium. In this paper, we report the transduction in *Salmonella* E group with phage ϵ_{15} (Iseki and Sakai, 1953) and ϵ_{34} (Harada, 1956a, b; Uetake, 1957).

MATERIALS AND METHODS

Media. Liquid cultures were prepared in Brain Heart Infusion (BHI) broth (Difco). The nutrient agar was Heart Infusion (HI) Agar. The medium AGGa agar consisted of medium A (Davis and Mingioli, 1950) containing 0.5% glucose, 0.5% sodium glutamate, and 1.3% Difco agar. The medium ALGa agar consisted of medium A containing 0.5% lactose (instead of glucose), 0.5% sodium glutamate, and 1.3% Difco agar.

Drugs. Dihydrostreptomycin (SM) sulfate, chloramphenicol (CM), tetracycline (TC) hydrochloride, and sulfisoxazole (SA) were used. Drug resistance was determined according to the modified method described by Fukumi (1953), scoring the maximal concentration of drugs diluted by twofold steps that allowed the growth of tested organisms.

Strains of bacteria and phages. *Salmonella newington* C_2(S-84), *S. chittagong* (S-224), phage ϵ_{15} (Iseki and Sakai, 1953), and phage ϵ_{34} (Harada, 1956a, b; Uetake, 1957) were used as transduction systems. The *Salmonella* strains used and their O-antigenic structures are shown in Table 1. *E. coli* K-12 strain PA 200 was supplied by B. D.

Fig. 2.6 The first discovery of the episome by Prof. Susumu Mitsuhashi

to the recipient cells. The (FR) factor was eliminated by treatment with acridine dye and also transduced. This paper, in the *Journal of Bacteriology* (Harada et al. 1963), was marvelous from not only a medical but also a genetic point of view.

Mitsuhashi added: "We were able to make such a discovery in Japan because we had many dysentery patients and tea parties with pickles contaminated with drug-resistant bacteria. In other words, if Japan had had good sanitation, such research would have been very difficult."

About 50 years ago, I chatted with him over afternoon tea with pickles at Gunma University.

Chapter 3
The University of Tokyo and Purdue University

Childhood and Early Schooling in Osaka, Ichikawa, Kumagaya, and Fudoka

Osaka

I was born on 28 October 1932 in the house of my mother Miyoshi's parents in Gunma Prefecture, as was common practice in those days. My father, Tomozou, graduated from Gunma Technical University and had worked for Shimazu Co. The company president, Genzou Shimazu, Jr., wanted to produce Japanese-made lead batteries because German batteries could not be imported during the First World War. The most serious problem was how to make lead powder for batteries and – so my mother told me – my father was of great assistance in producing this powder (Figs. 3.1 and 3.2).

I was a foolhardy boy and one day I fell into the river just behind my house. I was too young to swim, of course. My mother ran up, but she could not swim either. Fortunately a young man passing by jumped into the river and rescued me. If he had not been there, I should have drowned. I have no words to express my gratitude.

Ichikawa

We stayed in Osaka until 1940. My father was posted to Ichikawa, Chiba Prefecture, and I entered the second grade of Ichikawa Primary School. When I read aloud, my fellow students laughed at me. I realised that my Osaka dialect was incomprehensible to people in Chiba so gradually I learnt the Chiba dialect.

The next year, 1942, the Pacific War broke out and on April 18, 16 American B-25 bombers, captained by James Harold Doolittle, mounted the first air raid on the Tokyo area, including Ichikawa. I was in bed with a cold, and still remember the sound of machine guns (Fig. 3.3).

© Springer Japan 2016
K. Horikoshi, *Extremophiles*, DOI 10.1007/978-4-431-55408-0_3

Fig. 3.1 Koki with his
mother, Miyoshi, and father,
Tomozou

Fig. 3.2 Koki (2 years old)

In those days it was very difficult to find books for children because of the
shortage of paper, but my father's science books aroused my early interest in science.
A book by Professor Jun Ishiwara on the theory of relativity was one of them. Of
course, I could barely understand much of it, but the book gave an account of how
a number of scientists had gone to Africa to verify Einstein's theory by watching the

Fig. 3.3 Koki, sister Mieko, and parents

total eclipse of the sun and their data indicated that the theory was correct. That, I felt, was very romantic indeed.

Another book that seized my imagination was *History of Science for Boys and Girls*, also written by Ishiwara. He said that it was not true that Newton discovered the law of gravitation by watching an apple fall. Many people had seen apples fall from trees. So how did he discover it? Newton had spent many years studying a great variety of subjects, until, as a result, the law of gravitation flashed into his mind. When I read this chapter, it seemed to me that this was entirely different from what I had been taught at school. Could I trust my schoolmasters?

When I was 10 years old, my father found me reading one of his books on how to build a radio, so he bought me an old radio set which was falling apart. He said: "This is a wiring diagram; you should connect the wires up and build a radio." I assembled the radio as shown in the diagram, but it did not work. All that came out of the speaker was a loud noise. I went to the radio shop where he had bought the old radio and asked what was wrong. The owner showed me that the connection of the low-frequency transformer was incorrect and consequently it was oscillating. Once I had reconnected it correctly, that radio provided me with hours of good music. I was given the nickname Radio Boy.

A Fateful Encounter with Kazuo Hori

In 1943 my class teacher, Mr. Kazuo Hori, inculcated in me a deeper interest in science. At the time I was a victim of bullying and Mr. Hori told me: "You are good at science; try to use this talent to win." On 30 November 1944 Mr. Hori visited our

Fig. 3.4 Mr. Kazuo Hori and friends in 1943

house and his discussion with my father turned to the topic of the Jews. Suddenly he turned to me and said: "The Jews are really superb, top-class people in the fields of sciences and arts, even though they do not have their own nation. Japan is a very small country, you should be a top-level scientist like a Jew, like, for instance, Einstein."

Two days later, on December 2, he wrote on a square of fancy cardboard a poem by the Chinese poet Wei Zheng (AD 580–643): "It is what is in one's heart that moves others" (Figs. 3.4 and 3.5). His remark determined the direction of my life.

The following year we had 2 days of tests in all our subjects. Mr. Hori told me I was in the top rank. At the time I did not understand the purpose of the tests, but someone told me later that the Japanese Government had a specific programme to teach science to students who achieved the highest level, but the war meant that the project was abandoned.

Evacuation and Air Raids

In 1945 the war was getting worse and worse for Japan and we were evacuated to Kumagaya, Saitama Prefecture.

At about 11:00 PM on August 14, just 1 day before the war ended, my mother awoke suddenly and said: "Let's go to next door's air-raid shelter." Thirty minutes later our area was firebombed by 89 Boeing B-29 bombers. Our house was on fire and the noise was deafening. We supposed that a firebomb had fallen near by and fragments had burst into the house. All the window glass facing the road was shattered and pieces of napalm were burning all over the floor. The first thing to do

Fig. 3.5 Mr. Hori and Koki (*left*); 1944 poem by Wei Zheng (*right*)

was to extinguish the flames. It was only after we had finished fighting the fires that we realised that downtown Kumagaya had been almost completely burnt out (50 % of the houses). I was too young to feel any fear but it took me a long time to calm down.

The following morning I went to high school on foot. In the road I saw the corpses of about 300 people burnt to death. The sight was overwhelming (Fig. 3.6).

When I came home, my main concern was to find out where the bomb that had set our house on fire had fallen. I looked everywhere I thought it might be but I could not find it. There was only one place left. The day before I had begun to dig a shelter near the road but had given up because of the high water table. A napalm bomb had fallen just where the shelter-to-be was and had exploded. The M-47 bomb was 30 cm in diameter and 100 cm in length. I collected several fragments from the shelter. If I had been able to finish the shelter, the whole of my family would have been killed.

I have no memory of my parents' or neighbours' reactions to the raid, but I remember that I cooked rice using napalm oil.

On the way back to my home, I found a newspaper which reported the end of the war. However, my interest was only in the two burnt-out houses next door and the fragments of the firebomb.

Fig. 3.6 Burnt-out Kumagaya city

People's sense of values changed a great deal after the end of the war. Some adults said Japan should become one of the American states. I could not understand this.

I started to build another short-wave radio and before long I was able to listen to the Japanese news from Hawaii. A year or two later, I used the largest bomb fragment in my collection to fashion a frying pan because of the shortage of kitchen tools.

In February 1946 my father went to Osaka, but on the return journey he contracted typhoid. Typhoid was widespread at the time because of inadequate sanitation, and the disease killed a large number of people. Although he returned to Kumagaya, he passed away on February 20. I was too young to really comprehend my father's death. The eels that he brought us as a souvenir of Hamanako remain engraved on my memory.

Fudoka High School

We moved from Kumagaya to Hanyu after my father died. Fudoka High School, to which I transferred, is one of the oldest high schools in Saitama Prefecture. However, for me an old school did not mean a good school. I was too intent on building radios to get on with my studies. On the way to the high school, there was a radio shop that

Fig. 3.7 Homemade
short-wave radio

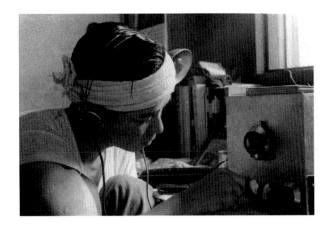

I visited every day. The owner was very skilful and made almost all the radios himself. From him I learnt how to make radio parts, and I built myself a short-wave radio using homemade parts, so I was able to enjoy listening to American stations such as HCJB in Honolulu (Fig. 3.7).

Discovery of Mould-Lysing Bacteria: The University of Tokyo and Purdue University

The University of Tokyo

In 1952 I entered the University of Tokyo to study mathematics, physics, chemistry, etc., for 4 years. In 1956 I joined the chemistry department of the university as a graduate student.

My major professor was Kin-ichiro Sakaguchi (Fig. 3.8). I chose Sakaguchi's laboratory because in 1955 I had built an oscillator to measure the dielectric constants of an alcohol and water mixture in his lab. During that time I had come into contact with many great scientists there (Fig. 3.9).

In 1956 Prof. Sakaguchi gave me the research theme for my doctoral thesis, "Autolysis of *Aspergillus oryzae*." He believed that it was an autolysate of *Aspergillus oryzae* that gave sake its flavour. Day after day I cultured stock strains of *Asp. oryzae*. After a week, I was required to taste the cultured fluid by *bero*-meter (*bero* means tongue in Japanese). This method was almost the same as that of sake testing (Fig. 3.10; sake-testing cup). I wished I could have used "modern instruments" to gather data, but I could not. I was young and felt that such an old-fashioned method was unscientific. However, the professor's words were like holy writ: I had to obey them if I wanted to receive my doctor's degree.

Fig. 3.8 Professor
Kin-ichiro Sakaguchi

Fig. 3.9 Members of Sakaguchi Laboratory: *arrow* indicates Koki

Fig. 3.10 Sake-testing
cup

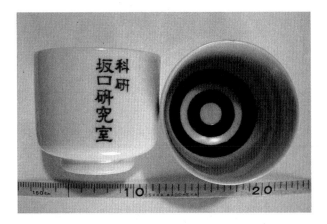

Fig. 3.11 *Left*: Culture flask in which *Aspergillus oryzae* is growing. *Right*: Culture flask from which *Asp. oryzae* has disappeared

Isolation of Fungi-Lysing Bacteria

One day in November I found one cultivation flask in which the mycelia of *Asp. oryzae* had completely disappeared (Fig. 3.11). The previous night when I had inspected the flasks, the mould had been flourishing in all of them. I remembered the spectacular pictures I had seen of how bacteria thrived and moved (Fig. 3.12; and Fig. 1 in the Preface). Yet no mycelium could be seen under the microscope.

The microorganism that was isolated from the flask was *Bacillus circulans* IAM1165. This culture fluid lysed *Asp. oryzae*. It was the first time that mould cells had been lysed by bacteria, and these results were published in *Nature* (Fig. 3.13; and Fig. 2 in the Preface). I took the photograph of the cell walls of *Asp. oryzae* that

Fig. 3.12 Contaminated bacteria

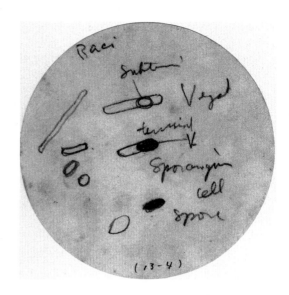

We thank Prof. K. Sakaguchi, Dr. M. Amaha and Dr. M. Nomura (University of Tokyo) for their encouragement and guidance.

KOKI HORIKOSHI
SIGEJI IIDA

Department of Agricultural Chemistry,
College of Agriculture,
University of Tokyo,
Bunkyo-ku, Tokyo,
and
Scientific Research Institute, Ltd.,
Bunkyo-ku, Tokyo.
Nov. 5.

[1] McCarty, M., J. Exp. Med., **96**, 555 (1952). Halvorson, H. O., and Greenberg, R. A., J. Bact., **69**, 45 (1955).
[2] Chung, C. W., and Nickerson, W. J., J. Biol. Chem., **208**, 395 (1954).
[3] Moore, S., and Stein, W., J. Biol. Chem., **211**, 909 (1954).

Printed in Great Britain by Fisher, Knight & Co., Ltd., St Albans

(Reprinted from Nature, Vol. 181, pp. 917–918, March 29, 1958)

Lysis of Fungal Mycelia by Bacterial Enzymes

MANY investigations have been published[1] on lysis of bacterial cells by microbial enzymes, but there seems to be few references to the lysis of fungal mycelia by microbial enzymes. A bacterial strain which caused visible lysis of *Aspergillus oryzae* mycelia has been isolated from soil. This strain belongs to the *Bacillus circulans* species ; it is Gram-negative, with spore-forming rods (spore : oval, terminal to subterminal), motile with peritrichous flagella, it reduces nitrate, is negative to acetyl-methylcarbinol, and produces acid from glucose without evolving gas.

When this strain was inoculated into a flask culture of *A. oryzae* (in Czapek–Dox medium), which had been previously shaken for 24 hr. at 30° C., almost all the mould mycelia disappeared after 80 hr. shaking at 30° C. The culture fluid was centrifuged at 7,000*g* for 30 min. to remove bacterial cells grown in the mixed culture, and a pale brownish supernatant was obtained.

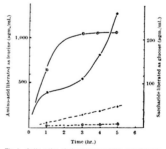

Fig. 1. Lytic action of the enzyme on living mycelial cells. 1 gm. of the living mycelial cells was suspended in 20 ml. of the lytic enzyme solution, and the mixture was incubated at 37° C. ●—●, Amino-acid; ○—○, saccharide; ●---●, amino-acid (control); ○---○, saccharide (control)

Fig. 3.13 Paper reported in *Nature*

appeared in my paper myself because the whole of the University of Tokyo had only one electron microscope.

I tried to lyse the cell wall 100 % but an insoluble residue still remained after a 1-week reaction. This insoluble material exhibited a fibrous structure under the electron microscope. The insoluble material could be hydrolyzed by HCl, and glucosamine was detected. Someone in the lab said to me: "Koki, it must be chitin." Chitin is a tough, pliable, widely distributed natural product resembling cellulose; its fibrous structure is derived from long unbranched chains of *N*-acetylglucosamine units connected by β-1,4-glucosidic linkages. He suggested to me that I should add chitinase in the reaction system. However, the question was: Who had available chitinase?

I carefully checked *Chemical Abstracts* and found one paper titled "Exocellular Chitinase from *Streptomyces* sp.", reported by D.M. Reynolds of the Department of Bacteriology at the University of California. He kindly sent me the enzyme sample.

The mixture of chitinase and my lytic enzyme was able to dissolve completely the cell walls of *Asp. oryzae*. I sent a manuscript of this result to *Nature* again before leaving Tokyo. The paper was published in 1959 (Fig. 3.14).

However, this bacterium showed very poor growth in the absence of mycelia of *Asp. oryzae* and the production of mould-lysing activity was very low. Therefore, it was clear that the purification of the lytic enzyme could only be done in a culture fluid containing mycelia of *Asp. oryzae*.

Purdue University

The first paper in *Nature* (Horikoshi and Iida 1958, 1959) changed my life. I received hundreds of letters from all over the world. One of them was from Professor Henry Koffler of Purdue University, Indiana. He invited me to join his research into the cell-surface structure of *Penicillium chrysogenum*, and he wished to use my *Asp. oryzae*-lysing *Bacillus circulans*. I accepted his invitation and decided to travel to West Lafayette, Indiana, by air. Fortunately, Fulbright funding sponsored my travel expenses (Figs. 3.15 and 3.16). On August 15, I arrived at West Lafayette via San Francisco (Fig. 3.17) and the Davis campus of the University of California. Henry, however, was in Switzerland, and the lab was in the process of moving to the brand-new Life Sciences Building. All I could do was wait for his return. Three weeks later we met Henry and his wife Phyllis at the university airport (Fig. 3.18). Someone else was also there, a man named Hubert Gottschling, from the University of Heidelberg, West Germany. Before long we became close friends and it was he who opened up European culture to me. If had not met him, I might not have visited Florence in 1968. In all probability I would not have discovered alkaliphilic microorganisms.

(*Reprinted from Nature, Vol.* 183, *pp.* 186-187, *Jan.* 17, 1959)

Effect of Lytic Enzyme from *Bacillus circulans* and Chitinase from *Streptomyces* sp. on *Aspergillus oryzae*

THE enzyme contained in a culture fluid of *Bacillus circulans* has been shown to exert lytic activity towards *Aspergillus oryzae* and that a polymer of melibiose was liberated from the cell wall. However, the cell wall was not completely lysed by the enzyme owing to the large amount of chitin present. During the search for enzymes causing lysis of fungal cells, it was found that a mixed preparation of chitinase and lytic enzyme exerted strong lytic activity on the cell wall of *A. oryzae*.

In this communication some observations pertaining to the action of chitinase and lytic enzyme, and to the fine structure of the cell wall are given.

The cell wall and the lytic enzyme were prepared by methods previously described. The crude chitinase was used as a chitinase preparation[2]. The mycelia of *A. oryzae* which had been previously shaken in

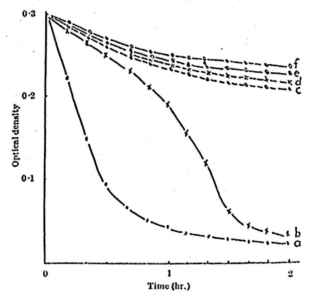

Fig. 1. Action of lytic enzyme and chitinase on the cell wall and on intact cells. Washed intact cells and cell wall were suspended in enzyme solutions (*tris*-buffer *M*/50, pH 6·8), and were incubated at 37° C. *a*, Cell wall (enzymes: lytic enzyme + chitinase); *b*, intact cells (enzymes: same as *a*); *c*, cell wall (enzyme: chitinase); *d*, cell wall (enzyme: lytic); *e*, intact cells (enzyme: lytic); *f*, intact cells (enzyme: chitinase)

Fig. 3.14 Paper published in 1959 in *Nature*

Fig. 3.15 Passport (1958)

Fig. 3.16 Passport (1958)

Fig. 3.17 At San Francisco

Fig. 3.18 Prof. Henry
Koffler and his wife Phyllis
(1958)

Fig. 3.19 In the laboratory

Endo-1,3-β-Glucanase

At Purdue I started once more to produce the mould-lysing enzyme *B. circulans* (Fig. 3.19). A Ph.D. candidate in our research room, John Greenawalt, was studying the cell surface of *Penicillium chrysogenum* and operated a Philips electron micro-scope in the lab by himself. I had the chance to give a lecture on my research into the enzyme *B. circulans* that had been published in *Nature*. The ensuing discussion was very serious, and some of the listeners did not believe in my purification methods. They said I did not know anything about the substrate itself because the cell surface was too complicated. Finally they suggested that I should purify more and determine the substrate specificity of the enzyme. Theoretically it was true, but in Japan at that time we did not have access to high-quality analytic instrumentation. The United States is a gladiatorial arena for science, in which many warriors fight, and a large audience watches to see which is the strongest. Nothing like this happened in Japan, and so it was that I learnt that science itself was a serious business and could be hard.

The first thing I had to find was the substrate specificity, but the purified enzyme could not hydrolyse any of the available polysaccharides. In March 1959 John gave me a polysaccharide, laminarin (1,3-β-glucan), that he had obtained from a friend in Norway. The enzyme hydrolysed the polysaccharide, and laminaribiose was detected in the hydrolyzate. This result indicated that the mould-lysing enzyme was endo-1,3-β-glucanase. I was so happy to have found the substrate specificity that I could not sleep. The next morning I told John about the previous night's results and I still remember his warm congratulations.

I then attempted to produce endo-1,3-β-glucanase in conventional media containing various nutrients. I noticed that the culture broth, which was neutral in pH, changed to an alkaline pH value. The addition of 0.5–1 % sodium bicarbonate to the culture broth gave good growth and production of the mould-lysing enzyme. The autolysis of *Asp. oryzae* changed the culture medium from weakly acidic to alkaline (pH 8–9). I found that such a change in pH value accelerated bacterial growth and endo-1,3-β-glucanase production. Of course, at that time we did not have the term "alkaliphiles." In 1975, I reconfirmed that this strain was also an alkaliphile which grew well in the alkaline culture media (pH 8.5–9) and produced lytic enzyme.

Seminars in Purdue

I attended many seminars during my 2-year stay, none of which I could have heard in Japan in the 1950s. The following seminars in particular impressed me.

1. *Discovery of lysozyme*

    In 1959 Prof. Koffler delivered a very stimulating lecture about lysozyme, an antibacterial enzyme widely distributed in body fluids and secretions such as tears and saliva. This enzyme hydrolyses the polysaccharide components of bacterial cell walls. Also, this enzyme was the first to prove that enzymes are proteins because they contain an amino acid sequence.

    My notebook tells the story of the discovery of the first enzyme.

    1921 was an important year. On 21 November Alexander Fleming's research notes record a great discovery, his natural antiseptic lysozyme. He had been busy growing a wide variety of microorganisms and observing how they behaved, in particular, their reactions to different substances. Fleming's bench was crowded with bacterial growths (known as cultures) for many weeks. He liked to leave them for a while and then have a good long look at everything before he discarded them. One never knew when something interesting might happen.

    One day, as he was looking through the piles of old culture dishes, he suddenly stopped. He peered at one carefully for several minutes. "This is interesting," he remarked.

    Fleming had been suffering from a cold some weeks earlier, and in his never-ending search to understand bacteria and disease, had started to grow a blob of his own thick nasal mucus. This dish was the one that had seized his attention. There were golden-yellow bacteria growing on the dish everywhere except around the blob of nasal fluid. They had started growing near by but had become glassy and seemed to be dissolving. And yet a little way from the nasal fluid they were growing normally. Was there something in the nasal fluid that actually killed bacteria? And did it happen with other people's nasal fluid? What about other body fluids? He tested saliva, pus, blood serum, and egg white. All had this amazing ability to stop these golden-yellow bacteria from growing.

    He published a number of reports about the lysozyme, but no one really grasped the significance of his discovery.

2. *Structure of DNA*

    In Japan in those days there was no chance at all of hearing seminars by famous scientists. But here in Purdue we had the chance to hear seminars after

Fig. 3.20 *From left:*
Professor Roy Doi
and Henry Halvorson

dinner almost every day. One of them was by James Watson, who gave us a lecture on the double helix of DNA. Not only biologists but also people from every field, with their wives, crowded into the large conference room. This was in 1959, before he received the Nobel Prize.

3. *Operon*

Another molecular biologist who visited us was Francois Jacob, who talked about the *Lac* operon. After his presentation, he mentioned that he did not do much in the way of biological experiments, but instead performed them solely in his brain. I was not sure whether I believed him. He also received the Nobel Prize in 1965.

4. *Bacterial spores*

Another visiting microbiologist was Roy H. Doi, a Japanese American. He worked on spores of *Bacillus subtilis* in Prof. Henry O. Halvorson's laboratory at the University of Wisconsin (Fig. 3.20). In 1966 he invited me as a post-doctorate to Davis, California. He is a very brilliant professor and was appointed a member of the American Academy of Science. We shall meet him again in a later part of this chapter.

Journey to Europe

Four years had passed since I had found the lytic phenomenon of *Asp. oryzae* and had shown that 1,3-β-glucan was one of the major components of the cell walls of *Asp. oryzae*. I imagined this marked the end of one chapter in my Ph.D. thesis. I decided I needed a change from microbiology. Suddenly I conceived a desire to visit Europe. I felt that if I returned to Japan, I would probably not be able to go to Europe. I asked Hubert about the best places to go.

Travel to Europe

I had a holiday allowance of 50 days from 7 January 1960. Through a travel agency near the university, I booked a flight with the cheapest airline – Icelandic Airways – and railway tickets. The total cost was about $500.

I went to New York via Indianapolis and took a plane to London from Idlewild Airport (now JF Kennedy Airport) on 10 January. The plane, a Douglas DC-4 (Fig. 3.21), was completely full. The person sitting next to me was English, and I told him I was visiting London just for sightseeing. The plane refuelled at Reykjavik, the Icelandic capital, where I walked around the cold, dark airport.

London

In 12 January 1960 I arrived at London Airport and reserved a room at a bed-and-breakfast near the British Museum. The room was very cold. London was very different from the USA. I paid two pence to turn on the gas heater. But it was too cold to sleep and I had to put on all the clothes that I had brought from Purdue.

The next morning I started sightseeing: the opera house in Covent Garden (I forget which opera I saw), Titian at the National Gallery, the Rosetta Stone in the British Museum. The following day I visited Buckingham Palace, Selfridges department store, and Piccadilly Circus (Fig. 3.22) by myself. There were only a few people out walking and I could not find a sightseeing bus. The weather was freezing cold. I took a train to Cambridge, although I had no appointments. I walked into Trinity College (Figs. 3.23 and 3.24) and saw the statue of Isaac Newton, which touched me to the core. "I am in England, not in the USA," I told myself excitedly, but I was extremely tired and so my memory of that time is hazy. However, I remember one thing very clearly: the manager of the B&B told me "You should not say

Fig. 3.21 Douglas DC-4
Icelandic Airways

Fig. 3.22 Piccadilly Circus

Fig. 3.23 Trinity College

Fig. 3.24 Near Trinity College

"yah" instead of "yes" here in London. I don't like hearing German." I greatly enjoyed my first visit to London.

Bruges, Belgium, and West Germany

On 16 January I took the boat-train from Victoria Station (Fig. 3.25) via Ostend to Bruges in Belgium. At the station, the stationmaster kindly made a reservation for me at a very nice hotel in town and I went sightseeing although the evening was already drawing in. The prevalence of statues of Christ on the street corners served as a joyous reminder that I really was in Europe. Suddenly, I heard the sound of a piano. Someone was playing Beethoven's *Moonlight Sonata*. It left an impression that I recall to this day.

The next morning I moved on via Brussels to Aachen in West Germany. Hubert had told me that the Holy Roman Emperors were crowned in the cathedral there and I was very surprised to find how small it was. It still bore signs of war damage and was very chilly.

From there I took the night train to Hamburg to meet Georg, Hubert's brother. He recognised me at the station without any trouble, as mine was the only Oriental face. He was a student at Hamburg University and lived near the Stadt Park underground station. I had heard that Hamburg had been largely destroyed by large-scale air raids during the war, but there were still many old houses in the park. I stayed with him for 2 or 3 days while I explored the city. I was especially pleased to see Paul Klee's painting *The Gate* in the Hamburg Museum.

Fig. 3.25 Victoria station (*golden arrow* shows counter)

Fig. 3.26 Family of Hubert Gottschling in Freden, Germany

Georg took me to Freden to meet Hubert's family (Fig. 3.26). It was a very tiny station; Georg and I were the only passengers to alight there. The family was German and had fled from Poland at the end of war. I was welcomed warmly and his mother baked a delicious cake for me. I spent a peaceful time in their house and I still remember with warmth the family and the German meals they served me. When I visited their family again in 1977 with my wife Sachiko and my son Toshiaki, Hubert's mother very kindly made the same cake for us.

Fig. 3.27   Venice

After Freden, I made my way via Göttingen University, Ulm, Münich, Zürich, and Vienna to Venice.

Italy

On 31 January, after crossing the lagoon, the night train brought me to Venice (Fig. 3.27) in the morning. I reserved a hotel a few minutes' walk from the station and walked around visiting cathedrals, basilicas, and museums. I also enjoyed delicious meals such as spaghetti and fried fish. Before coming to Venice, I did not know that pasta was a starter and a main dish was to follow. I was just a country boy from Indiana. I was particularly impressed by the Basilica di Santa Maria Gloriosa dei Firari. Titian's famous masterpieces, *The Assumption of the Virgin* and *Pesaro Madonna*, were beyond words.

After 4 days in Venice, I took a train to Milan. Those were the days before da Vinci's *Last Supper* had been restored and I remember the mural as being very faint and dark. A photograph on the wall showed how the fresco was protected by sand-bags during the war. I had a chance to see an opera in the Teatro alla Scala (Fig. 3.28), but I forget the name.

The next place on my itinerary was Florence. I spent 3 days visiting the Uffizi Museum. It was so dark and cold in wintertime, and the entrance gate was partially closed. But once I had opened the door the inside was beautiful and silent. On my last day there was a special Titian exhibition at the Pitti Palace. The *Penitent Magdalene* was unforgettable. I promptly resolved to see all Titian's paintings, although to date my quest is still unfinished. However, it would not be until 1968 that I was able to revisit Florence (see Chap. 4), where I was to have something of an epiphany.

Fig. 3.28 Teatro alla Scala in Milano

In Rome I had the chance to see Verdi's *La Forza del Destino* at the opera house. That was my first time to see it, although I had heard the music many times on recordings. It was a most exciting experience.

I visited the Coliseum (Fig. 3.29), the Roman Forum, the Trevi Fountain, and all the many sights detailed in the travel guide that I had bought in the USA. There were only a few other tourists walking around. It was nothing like the film *Roman Holiday* starring Audrey Hepburn, but it was my own personal winter Roman holiday (Fig. 3.30). History and reality blended together in beautiful harmony.

I travelled on to Naples and watched *Pagliacci* and *Cavalleria Rusticana* at the opera house in somewhat modernised versions. The next day I visited Pompeii. There were very few visitors and neither gate nor entrance fee. In the peace and quiet I was happy to walk the paved streets, to visit the *House of the Faun*, to admire the portrait of the baker Terentius, and to explore the Temple of Jupiter. In the distance I could see Mount Vesuvius (Fig. 3.31) without any smoke rising. Such a peaceful mountain it seemed.

Paris

From Naples I made my way by train via Milan and Zurich to Paris.

My first port of call was the Louvre, to see the Venus de Milo, the *Mona Lisa*, and paintings by Titian. The museum was not crowded, and as I went for a walk beside the Seine I was delighted with the thought that here I was in Paris (Figs. 3.32, 3.33 and 3.34).

Fig. 3.29 Coliseum in Rome

Fig. 3.30 Winter Roman holiday. No tourists!

Fig. 3.31 Pompeii and Mount Vesuvius. Very quiet and no visitors

Fig. 3.32 Paris, the Eiffel
Tower

Fig. 3.33 Paris

Fig. 3.34 Paris

I strolled along the Champs Élysées, and found a cinema that was showing Charlie Chaplin's *The Great Dictator*. A notice announced that the star himself was expected to attend so I entered, hoping to see him, but the place was already too crowded.

My four days in the city passed quickly and I took a train to Amsterdam on February 19. Before my flight back to the USA, I was able to see Rembrandt's *The Night Watch* in the Rijksmuseum. The Icelandic Airlines flight plane left Schiphol Airport for New York, and I returned to Purdue on February 23 (Fig. 3.35).

"Adieu Europe and a Marvellous Youth!"

Return to Japan

I made plans to return to Tokyo in the middle of June. The first thing I had to do was to make the purified β-1,3-glucanase enzyme. I had various methods from which to choose.

If I had had a large amount of laminarin, I could cultivate *Bacillus circulans* in a medium containing laminarin, and thus produce the crude enzyme. But I had only 10 g.

I could produce large amounts of the enzyme if I cultivated it in a living *Asp. oryzae* medium, but much of it was contaminated by the presence of other enzymes.

If I used a medium containing heat-killed *Asp. oryzae*, I could produce the enzyme, but only in very small amounts. Suddenly, I remembered that the pH value of method no. 2 was 8–9 and gave off a smell similar to ammonia.

I increased the pH value to about 9 by adding sodium bicarbonate. Large amounts of the crude enzyme were thus produced. I was really surprised by the effect of the changed pH value. Now I would be able to make the purified enzyme in Tokyo.

Fig. 3.35 Back to Purdue from Europe

Fig. 3.36 Ford 1945 (only $25!)

The second thing I had to do was to sell my used 1945 Ford (Fig. 3.36). The price was only $25, and I bought a new Parker fountain pen for the same price. A portable typewriter, LP records of classical music, instant coffee, and a pair of skiing boots bought in Italy were in my luggage.

To Hanyu via San Francisco

On 31 May I began the long journey from Purdue to San Francisco by train. It took three days and two nights. The Great Salt Lake in the glow of the setting sun was impressive. I booked flights from San Francisco to Los Angeles by DC-8 and from Los Angeles to Tokyo by DC-6. Then, I sent my mother a telegram through Western Union.

On June 4, as the plane took off from its stopover in Honolulu, the captain announced: "Because of trouble with one of the engines, two of the propellers have to be shut off for the sake of balance." Our four-engined plane had suddenly become a twin-engine.

We landed at Midway Airport to fix the faulty engine. There were several sunken ships and anti-aircraft guns from the war near the runway. I collected some white corals as a souvenir.

On June 6 the plane arrived safely in Tokyo and I returned home. My mother was surprised to see me and asked: "Are you real? When did you get back?" My telegram from Los Angeles had not reached her.

My youth and 2 years' study abroad were over. I was exhausted. The only thing I wanted to do was to sleep in my own bed.

Return to the University of Tokyo and Marriage

On July 1 I returned to the chemistry department of the University of Tokyo to continue my Ph.D. thesis, *Lysis of Aspergillus oryzae by Bacillus circulans*. My major professor was Kei Arima (Fig. 3.37), the successor to Kin-Ichiro Sakaguchi. In his laboratory, a new drug for athlete's foot was discovered that was put into industrial production.

Fig. 3.37 Koki, Roy Doi, and Prof. Kei Arima

For ethical reasons many vegetarians would not eat cheese solidified by calf rennet. Arima isolated microbial rennet from *Mucor* mould and thus allowed cheese to be produced for vegetarians.

I purified β-1,3-glucanase in the lytic enzyme made in Purdue by DEAE-cellulose. The cell wall of *Asp. oryzae* was hydrolysed by the purified enzyme. The insoluble residue was chitin, which was confirmed by X-ray analysis. A mixture of β-1,3-glucanase and chitinase could solubilize the cell wall of *Asp. oryzae*. The mixed enzymes, furthermore, could lyse *Penicillium chrysogenum* and *Saccharomyces cerevisiae*. These results indicated that the three cell walls tested had the same structures containing β-1,3-glucanase and chitin (Fig. 3.38).

With this research I finished at The Tokyo University's graduate school of chemistry in 1961 and entered the Institute for Physical and Chemical Research (RIKEN).

Marriage

The most important thing in my life came about as the result of a telephone call from one of my friends, Toshihiko Tsukamoto. He was brewing wine in Yamanashi Prefecture, about 100 miles from Tokyo. He said, "I know someone I would like to introduce to you. Meet me at Akasaka underground station." At the station he introduced me to a young lady who was working for one of the foreign companies in Aoyama. This was my first meeting with my future wife, Sachiko (Fig. 3.39). We chatted about all sorts of things in a coffee shop. She said, "The English language is very difficult to speak fluently, didn't you agree?" "Yes, indeed I do. I cannot speak good English!"

Her father Shigeo (Fig. 3.40) had been in London, and her father and mother (Shizuko) had lived in Buenos Aires in the early years of the twentieth century.

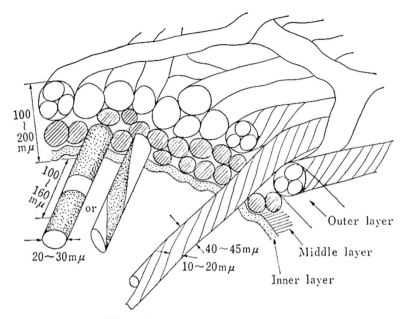

100~200 mμ

100~160 mμ

20~30 mμ

or

40~45 mμ

10~20 mμ

Outer layer

Middle layer

Inner layer

Fine Structure of the Cell Wall of
Asp. oryzae.
Outer layer : Polysaccharide fiber
Middle layer : Chitin fiber
Inner layer : Polysaccharide

Fig. 3.38   Fine structure of *Aspergillus oryzae*

Fig. 3.39   Sachiko Hamada

Fig. 3.40 Shigeo Hamada at Chiswick, London (*left*); Shizuko Hamada at Buenos Aires, Argentina (*right*)

Sachiko's two brothers were born in Buenos Aires, although she was "made in Japan." I had never had such cosmopolitan friends. I visited her house often and loved to hear from her father about his travels in the UK in 1912. Sachiko's elder brother had died in the Second World War and her younger brother was a microbiologist. I was very happy to talk with someone of the same generation as my late father.

In 1963 I proposed marriage to Sachiko in the presence of her parents and we were married on 14 November, her birthday. We went to the Imperial Hotel in Tokyo to stay the night, but it turned out that the hotel had no reservation in our name. Sachiko phoned her close friend who had undertaken to reserve a hotel room. She told her that the room was booked at a different hotel, the Ohkura. The next day we went to the Shiga ski resort for our honeymoon. On the way home, we stopped at my mother's house. Early in the morning of 23 November we were shocked to learn of the assassination of President John F. Kennedy. The news came via the first satellite transmission. How quickly information could spread.

Post Doctorate at Davis

On 29 September 1964 our son Toshiaki was born.

My research at RIKEN was greatly influenced by the work I had done in Purdue. The structure of the spore coat of *Asp. oryzae* and the germination of spores of the mould were published in the *Journal of Bacteriology* (Horikoshi and Iida 1964; Horikoshi et al. 1965; Horikoshi and Ikeda 1965). I was, however, shocked by Prof. Harlyn O. Halvorson, who told me *Aspergillus oryzae* was not in widespread use in the USA and moreover that I had just rehashed bacterial spore work we had already done. He was a big name in the field of spore research.

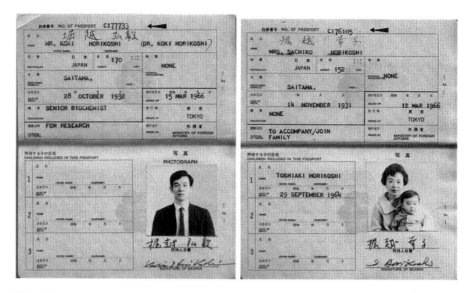

Fig. 3.41 Passports

I could not decide which field I was going to study. I asked Sachiko what I should do. Her answer was surprising but simple: "Let's go to the United States to learn a different way of thinking." I was delighted with this suggestion. I could not possibly put my gratitude into words.

I wrote a letter to Professor Roy Doi at the University of California at Davis. He kindly invited me to come as a post-doctorate student to work on protein synthesis of *Bacillus subtilis*. In the 1960s it was very hard for people in Japan to travel abroad because of the shortage of foreign currency. In April 1966 I was awarded the Prize for Agricultural Chemistry (for the discovery of mould-lysing enzymes). The prize money allowed us to book a Japan Airlines flight from Tokyo to San Francisco. I was able to get a researcher's visa relatively easily, but obtaining visas for Sachiko and Toshiaki was more complicated, They had to visit the U.S. Consular Office and swear an oath (Fig. 3.41).

On 9 May I arrived at our dormitory at 1–5 Orchard Park, Davis, just in front of a Safeway supermarket. On 1 June Sachiko and Toshiaki arrived at San Francisco airport. When Toshi saw me he cried out, "Papa, Papa," and jumped into my arms.

That evening we arrived at our house. They were delighted to find cantaloupes and oranges in the refrigerator, especially cantaloupes as they had never eaten them before.

This dormitory was very international and the next day Sachiko met our neighbour, who came from Norway. She asked Sachiko where she came from. When Sachiko answered that she had arrived from Japan the day before, the woman was surprised. "I thought you had come from another American state," she said.

Toshi quickly made a number of friends, one of whom was our neighbour's son, Douglas. Listening to the two of them together, I was amazed at how quickly Toshi picked up idiomatic English. Someone asked Toshi what his name was, to which he

Fig. 3.42 Toshi and friends at Davis

naturally replied "Toshi." But after a week or so, his answer had changed. "My name is Peter," he would say, or sometimes, "My name is Batman." I watched this change with interest. As the first month passed, to her embarrassment Sachiko found that she was known to the neighbours simply as "the mother of that happy boy Toshi." His language development was truly remarkable (Fig. 3.42).

Way of Thinking in English

I decided to learn English again from scratch. But what was the best way to study the language?

I bought a used car through Bob Sofue, a Japanese American. Almost every other day he came to our house with his girlfriend, Toshiko. One day, Sachiko went to Nevada with her and won $250 gambling. Beginner's luck!

We chatted about all sorts of things together, and found many differences between the English and Japanese languages (Figs. 3.43 and 3.44). When people bought petrol, I heard them say, "Fill'r up regular, please." Bob taught me that they actually said, "Fill her up regular, please." This was a good example of how to acquire practical American English. More importantly, I learnt a different way of thinking. For instance, if you want to negotiate a salary increase with your boss, do not give him the chance to say no. "Are you satisfied with my work?" you ask. To which he will answer yes.

LIFE IN JAPAN — Japanese women of the Davis Community church Chatter Group prepare the program at the Monday night meeting of the group. Seated from left are Mmes. Yoshito Goto, Reiko Masuzawa and Chieko Katayama. Standing from left are Mmes. Rumiko Matsumura, Sachiko Horikoshi, Ayako Inukai, Mariko Yagi, T a k a k o Iwahuri, Kimi Tamura and Grace Noda, program chairman. A movie, "Life in the Japanese Home," was shown and Japanese refreshments were served.

Fig. 3.43   Chatter group in Davis

Fig. 3.44   Gold Discovery Site in California with Toshiaki and Sachiko. We were able to get about 2 g of gold dust from the river

Follow this up with "I am very happy to work with such a good professor. And I would be even happier if you would be kind enough to give me a pay rise."

I told Sachiko this story and she said she had also had a similar talk with her boss in Union Carbide in Tokyo. In support of her request for a pay rise, she told him, "I can speak very posh Japanese, which no one else in our office can speak." Sometimes Roy Doi would say, "In California you seem like an American, but in Japan you are Japanese. How are you able to change your way of thinking?" It must have been the result of Bob's teaching.

The difference in the way of thinking can be seen in the difference in sentence structure. In Japanese we would say: "It is going to rain so I have brought an umbrella." In English: "I have brought an umbrella because it is going to rain." French, German, and Chinese, all have their own different ways of thinking, although they have the same end result. East is East and West is West.

The theme of the research in Roy's lab was protein synthesis in germinating spores of *Bacillus subtilis*. The C-terminals of proteins synthesised in a cell-free system were alanine, which was different from the *Escherichia coli* system (methionine is the C-terminal). This finding was reported at the Federation Meeting of Bacteriology in New York. The chairman was Seviro Ochoa, a Nobel Laureate. I received a great many questions, opinions, and suggestions. One was from Henry Koffler of Purdue University, who announced some very interesting results that were published in the *Journal of Biological Chemistry* (Horikoshi and Doi 1968).

Back to RIKEN

We left Davis in June 1967. When we arrived at Haneda Airport, Toshiaki hugged his safety blanket tightly and cried out: "Let's go back to Davis, here everyone has black hair." And Sachiko's mother teased us: "You look like refugees from the USA."

During our stay in Davis, RIKEN had moved from Komagome to Wako-city. For the time being, I was content to investigate the germination of spores of *Aspergillus oryzae*. Spores of *Aspergillus oryzae* are similar to seeds of plants and have reserve substances for germination. They just need water for germination and have the same protein-synthesising systems as those of vegetative cells (Horikoshi and Ikeda 1969; Horikoshi 1971a).

But what was I going to do for future research? I had no idea.

Chapter 4
Alkaliphiles

Discovery of Alkaliphiles

When Did I Start Investigating Alkaliphiles?

A serendipitous story in Florence in Italy caused me to begin research on alkaliphiles in 1968.

In 1967, after my return from Davis, I found myself at a dead end as far as research was concerned. I have an old passport with a black cover issued on 17 September 1968 (Fig. 4.1) and, using this, I left Haneda Airport on 1 October, and flew to Europe, in the hope that it would lift me out of the slump. I toured many of the places I had visited on my trip in 1960 and spent about a month just seeing the sights.

At the end of October, I visited Florence, Italy, and saw the Renaissance buildings, a style of architecture utterly different from that of Japan. Although both cultures flourished at the same period – between the fourteenth and fifteenth centuries – no Japanese at that time could have imagined this Renaissance flowering on the other side of the world (Fig. 4.2).

Then suddenly I heard a voice whispering in my ear, "There could be a whole new world of microorganisms in different unexplored cultures." Louis Pasteur used bouillon to cultivate microorganisms and Robert Koch cultivated bacteria on potatoes. Memories of the experiments on moulds lysed by *Bacillus circulans* that I had done almost 10 years before flashed into my mind. Could there be an entirely unknown domain of microorganisms existing at alkaline pH? The acidic environment was being studied, probably because most food is acidic. Little work had been done in the alkaline region. Almost all biologists believed that life could survive only within a very narrow range of temperature, pressure, acidity, alkalinity, salinity, and so on, in what were termed "moderate" environments.

So when microbiologists looked around for interesting bacteria and other life forms, they attempted to isolate microorganisms only from moderate environments. But science, just as much as the arts, relies upon a sense of romance and intuition.

© Springer Japan 2016
K. Horikoshi, *Extremophiles*, DOI 10.1007/978-4-431-55408-0_4

Fig. 4.1 Passport for alkaliphiles

Upon my return to Japan on 2 November, I prepared two alkaline media containing 1 % sodium carbonate, "Horikoshi-I and Horikoshi-II" (Table 4.1), put small amounts of soil collected from various areas within the Institute of Physical and Chemical Research (RIKEN) into 30 test tubes, and incubated them overnight at 37 °C. To my surprise, various microorganisms flourished in all 30 test tubes. I isolated a great number of alkaliphilic microorganisms and purified many alkaline enzymes.

Here was a new alkaline world that was utterly different from the neutral world discovered by Pasteur and Koch. I named these microorganisms, which thrive in alkaline environments, "alkaliphiles." This was my first encounter with alkaliphiles.

Fig. 4.2 Kinkakuji Temple in Kyoto and Duomo in Firenze (East is East and West is West!)

Table 4.1 Two basic media for alkaliphiles

	Horikoshi-I (g/l)	Horikoshi-II (g/l)
Glucose	10	–
Soluble starch	–	10
Polypeptone	5	5
Yeast extract	5	5
KH_2PO_4	1	1
$Mg_2SO_4 \cdot 7H_2O$	0.2	0.2
Na_2CO_3	10	10
Agar for plates	20	20

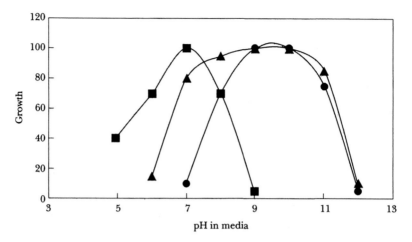

Fig. 4.3 Growth curves at different pH values. *Squares* neutrophilic, *triangles* alkalitolerant, *circles* alkaliphiles

Alkaliphilic Microorganisms

There are no precise definitions of what characterises an alkaliphilic or alkalitolerant organism. Several microorganisms exhibit more than one pH optimum for growth depending on the growth conditions, particularly nutrients, metal ions, and temperature. In this book, therefore, the term "alkaliphile" is used for microorganisms that grow optimally or very well at pH values above 9, often between 10 and 12, but cannot grow or grow only slowly at the near-neutral pH value of 6.5 (Fig. 4.3).

Distribution and Isolation of Alkaliphiles

Alkaliphilic microorganisms coexist with neutrophilic microorganisms, as well as occupying specific extreme environments in nature. Figure 4.4 illustrates the relationship between the occurrence of alkaliphilic microorganisms and the pH of the sample origin. Sodium carbonate is generally used to adjust the pH to around 10, because alkaliphiles usually require at least some sodium ions. Figure 4.5 shows a very small but very active research group: from the left, Koki, Mrs. Kurono, Mrs. Shishido, and Mr. Kurono.

The frequency of alkaliphilic microorganisms in neutral "ordinary" soil samples is 10^2–10^5/g of soil, which corresponds to 1/10–1/100 of the population of the neutrophilic microorganisms. Recent studies show that alkaliphilic bacteria have also been found in deep-sea sediments collected from depths as great as the 10,898 m

Fig. 4.4 Distribution of alkaliphiles in soils

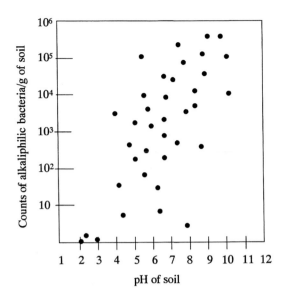

Fig. 4.5 A very small but very active research group: *from left*, Koki, Mrs. Kurono, Mrs. Shishido, and Mr. Kurono

of the Mariana Trench. Many different kinds of alkaliphilic microorganisms, including bacteria belonging to the genera *Bacillus*, *Micrococcus*, *Pseudomonas*, and *Streptomyces*, and eukaryotes such as yeasts and filamentous fungi, have been isolated from a variety of environments.

Physiological Features of Alkaliphiles
(Horikoshi and Akiba 1982; Horikoshi 1991)

Internal pH

Most alkaliphiles have an optimal growth pH of around 10, which is the most significant difference from the well-investigated neutrophilic microorganisms. Therefore, the question arises how these alkaliphilic microorganisms can grow in such an extreme environment.

Is there any difference in physiological and structural aspects between alkaliphilic and neutrophilic microorganisms? Internal cytoplasmic pH can be estimated from the optimal pH of intracellular enzymes. For example, α-galactosidase from an alkaliphile, *Micrococcus* sp. strain 31-2, has an optimal catalytic pH at 7.5, suggesting that the internal pH is around neutral. Furthermore, the cell-free protein synthesis systems from alkaliphiles optimally incorporate amino acids into protein at pH 8.2–8.5, only 0.5 pH unit higher than that of the neutrophilic *Bacillus subtilis*.

Another method to estimate internal pH is to measure, in cells, the inside and outside distribution of weak bases, which are not actively transported by cells. The internal pH was maintained at around 8, despite a high external pH of 8–11 (Fig. 4.6).

Intracellular Enzyme System

When I began to study the physiology of alkaliphiles in the 1970s, the properties of intracellular enzymes of alkaliphilic microorganisms were very attractive. How did extreme environments affect intracellular enzymes? Were there characteristic

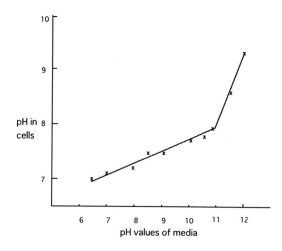

Fig. 4.6 pH values of media and intracellular pH values

Table 4.2 Properties of α-galactosidases

Property	Micrococcus sp. no. 31-2	Bacillus sp. no. 7-5	Mortierella vinacea
Molecular weight	367,000	312,000	
Optimum pH	7.5	6.5	4.0–6.0
pH stability range	7.5–8.0	6.0–8.5	7.0–11.0
Optimum temperature, °C	40	40	
K_m ONPG, mM	0.47	1.0	0.36±0.014
Meliboise, mM	1.5	7.9	0.39±0.029
Raffinose, mM	12.6	24.1	1.83±0.13

differences between intracellular enzymes of neutrophils and alkaliphiles? Many intracellular enzymes were isolated and purified, and their properties were investigated. Some enzymes exhibited relatively higher pH optima than neutrophilic bacteria. However, there appeared to be nothing exceptional in intracellular enzymes. Cell-free protein synthesis systems showed maximal activity between pH 8 and 8.5, but this was only 0.5 pH unit higher than that observed in neutrophilic *B. subtilis*.

α-Galactosidases and β-Galactosidases

Two kinds of α-galactosidase-producing bacteria, *Micrococcus* sp. no. 31-2 and *Bacillus* sp. no. 7-5, were isolated from soil on a modified Horikoshi-I medium (1 % raffinose was substituted for glucose). Alkaliphilic *Micrococcus* sp. no. 31-2 induced a cytoplasmic α-galactosidase and alkaliphilic *Bacillus* sp. no. 7-5 produced an extracellular α-galactosidase constitutively. The properties of these enzymes were similar (Table 4.2). The optimum pH range of these enzymes was higher than that of yeast, mould, and plant seeds.

β-Galactosidase was produced by alkaliphilic *Bacillus halodurans* C-125, not only in alkaline medium but also in neutral medium. However, the induction of the enzymes occurred much more rapidly at pH 10.2 than at pH 7.2. The enzymes produced in media of different pH values possessed the same enzymatic properties. The molecular weight of the enzyme was about 185,000. The enzyme was most active at pH 6.5 and stable over the pH range 5.5–9.0.

RNA Polymerases and Protein-Synthesizing System

Many alkaliphilic *Bacillus* strains have been isolated and their properties studied in the author's laboratory, but no critical differences were observed.

We published on a cell-free protein-synthesizing system using alkaliphilic *B. halodurans* A-59 and C-125. Microorganisms grown in Horikoshi-I medium

Fig. 4.7 Protein synthesis
directed by poly-U

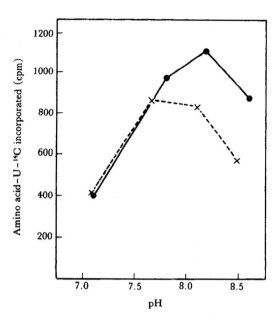

were collected by centrifugation (6,000 g) and washed twice with 0.01 M Tris-HCl
buffer (pH 7.5) containing 0.01 M MgCl$_2$, 6 mM 2-mercaptoethanol, and 0.06 M
KCl (TMM buffer). The enzymes were extracted with TMM buffer equivalent to
twice the volume of cells ground with two to three times their weight in alumina in
a prechilled mortar. The extract was centrifuged twice at 30,000 g for 30 min.
The upper two thirds of the supernatant was dialyzed against TMM buffer for 3 h.
From this supernatant fluid, ribosomes were precipitated by centrifugation at
105,000 g for 2 h, washed once with TMM buffer, and finally resuspended in
TMM buffer. The supernatant was centrifuged at 105,000 g for 2.5 h and the upper
two thirds used as the S-100 fraction. tRNA was prepared from the cells by the
conventional method.

The optimal pH for protein synthesis directed by poly-U or endogenous mRNA
was about 8.5, which was about 0.5 higher than that of *B. subtilis* (Fig. 4.7). As the
reference, the protein-synthesizing system of *B. subtilis* was tested under the same
condition. To confirm this result, protein synthesis directed by endogenous mRNA
was performed and the optimal pH was also slightly higher than that of *B. subtilis*
(about pH 8.5).

Ribosomes of alkaliphilic *B. halodurans* A-59 and C-125 were of the 1970s type,
and no difference was observed in their thermal denaturation curves. Phenylalanyl-
tRNA synthetase activity at different pH values also indicated no remarkable
difference between alkaliphilic *Bacillus* and *B. subtilis* Marburg 168. Phenylalanine
incorporation was tested in a series of homogeneous and heterogeneous combinations
(Table 4.3). In the systems containing alkaliphilic *Bacillus* ribosomes or S-100,

Table 4.3 Phenylalanyl-tRNA synthetase activities

Ribosomes	S-100	Phenylalanine incorporated (nmol)		Ratio between pH 8.4 and 7.5
		pH 7.5	pH 8.4	
Bacillus A-59	Bacillus A-59	5.82	9.37	1.61
	Bacillus C-125	6.12	8.65	1.41
	Bacillus subtilis	5.84	6.37	1.09
B. C-125	Bacillus A-59	5.17	6.62	1.28
	Bacillus C-125	5.32	8.12	1.53
	Bacillus subtilis	5.63	6.46	1.15
B. subtilis	Bacillus A-59	6.31	7.15	1.13
	Bacillus C-125	6.71	8.34	1.24
	Bacillus subtilis	9.65	6.59	0.68

No remarkable difference between alkaliphilic *Bacillus* and *B. subtilis* Marburg 168

phenylalanine incorporation was higher at pH 8.4 than at pH 7.5, although the activity of heterogeneous systems was lower than that of homogeneous systems. In conclusion, alkaliphilic *B. halodurans* A-59 grows well under highly alkaline conditions, but the protein-synthesizing mechanism is essentially the same as that of *B. subtilis*. The pH optimum of the protein-synthesizing mechanism strongly suggests that the internal pH value may be 8–8.5, not 10. Heterogeneous combination also supports this possibility. The results clearly indicate that the differences between alkaliphilic *Bacillus* and neutrophilic *Bacillus* exist on the cell surface and not within the cells.

Therefore, one of the key features in alkaliphily is associated with the cell surface, which discriminates and maintains the intracellular neutral environment separate from the extracellular alkaline environment.

Cell Walls (Horikoshi 1991, 1999a; Horikoshi and Grant 1998)

Acidic Polymers in the Cell Wall

Because the protoplasts of alkaliphilic *Bacillus* strains lose their stability in alkaline environments, it has been suggested that the cell wall may play a key role in protecting the cell from alkaline environments. Components of the cell walls of several alkaliphilic *Bacillus* sp. have been investigated in comparison with those of the neutrophilic *B. subtilis*. In addition to peptidoglycan, alkaliphilic *Bacillus* sp. contain certain acidic polymers, such as galacturonic acid, gluconic acid, glutamic acid, aspartic acid, and phosphoric acid. The negative charges on the acidic nonpeptidoglycan components may give the cell surface its ability to adsorb sodium and hydronium ions and repel hydroxide ions and, as a consequence, may assist cells to grow in alkaline environments.

Table 4.4 Cell wall composition of alkaliphilic *Bacillus* strains

Group	Components of cell walls	Growth pH	Ion requirement
1	High in glucuronic acid, teichuronic acid, and hexosamine	No growth at pH 7	Na⁺ (essential)
2	High in glutamic acid, aspartic acid, galacturonic acid, glucuronic acid, and teichuronic acid	Capable of growth at pH 7	Na⁺ (essential)
3	Presence of phosphoric acid, similar to *B. subtilis*	Capable of growth at pH 7 and 10 in the presence of Na⁺ or K⁺ (both are essential)	

Peptidoglycan Composition

The peptidoglycans of alkaliphilic *Bacillus* sp. appear to be similar to that of *B. subtilis*. However, their composition is characterised by an excess of hexosamines and amino acids in the cell walls compared to that of the neutrophilic *B. subtilis*. Glucosamine, muramic acid, D- and L-alanine, D-glutamic acid, *meso-diaminopimelic* acid, and acetic acid were found in the hydrolysate. Although some variation in the amide content among the peptidoglycans from alkaliphilic *Bacillus* strains was found, the variation in pattern was similar to that known in neutrophilic *Bacillus* species. These data are summarised in Table 4.4.

Flagella

Some alkaliphiles are motile by means of flagella. This flagella-induced motility is driven by a sodium- or potassium-motive force instead of the proton-motive force typical of neutrophils.

The flagellin protein gene of *Bacillus halodurans* C-125 was cloned and expressed in *E. coli*. This flagellin shares high homology with other known flagellins, such as that of *B. subtilis* 168.

Na⁺ Ions and Membrane Transport

Alkaliphilic microorganisms grow vigorously at pH 9–11 and require Na⁺ for growth. The presence of sodium ions in the surrounding environment has proved to be essential for effective solute transport through the membranes of alkaliphilic *Bacillus* sp. According to the chemiosmotic theory, the proton-motive force in the cells is generated by the electron transport chain or by excreted H⁺ derived from ATP metabolism by ATPase. H⁺ is then reincorporated into the cells with cotransport of various substrates.

In Na⁺-dependent transport systems, the H⁺ is exchanged with Na⁺ by Na⁺/H⁺ antiporter systems, thus generating a sodium-motive force, which drives substrates accompanied by Na⁺ into the cells. The incorporation of a test substrate, α-aminoisobutyrate, increased twofold as the external pH shifted from 7 to 9, and the presence of sodium ions significantly enhanced the incorporation; $0.2\ N$ NaCl produced an optimum incorporation rate that was 20 times the rate observed in the absence of NaCl. Other cations, including K^+, Li^+, NH_4^+, Cs^+, and Rb^+, showed no effect, nor did their counteranions.

Mechanisms of Cytoplasmic pH Regulation

The cells have two barriers to reduce pH values from 10.5 to 8 (Fig. 4.8). Cell walls containing acidic polymers function as a negatively charged matrix and may reduce the pH value at the cell surface (Tsujii 2002). The surface of the plasma membrane must presumably be kept below pH 9, because the plasma membrane is very unstable at alkaline pH values (pH 8.5–9.0) much below the pH optimum for growth.

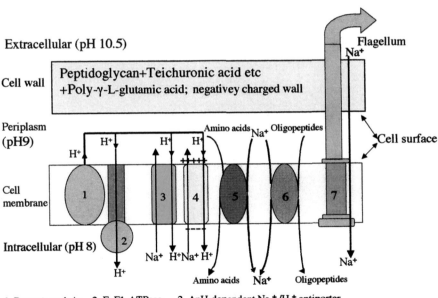

1. Respratory chain   2. FoF1-ATPase   3. ΔpH-dependent Na⁺/H⁺ antiporter
4. Δψ-dependent Na⁺/H⁺ antiporter   5. Amino acids/Na⁺ symporter
6. Oligopeptides/Na⁺ symporter   7. Flagella motor (mot)

Fig. 4.8 Cytoplasmic pH regulation

Plasma membranes may also maintain pH homeostasis by using the Na⁺/H⁺ antiporter system ($\Delta\psi$ dependent and ΔpH dependent), the K⁺/H⁺ antiporter, and ATPase-driven H⁺ expulsion. Recent studies on the critical antiporters in several laboratories have begun to clarify the number and characteristics of the porters that support active mechanisms of pH homeostasis.

Chromosomal DNA of Bacillus halodurans *C-125*

How alkaliphiles adapt to their alkaline environments is one of the most interesting and challenging topics that might be clarified by genome analysis. A physical map of the chromosome of *B. halodurans* C-125 has been constructed to facilitate whole-genome analysis (Takami et al. 1999).

In 2000, we reported the complete genome sequence of *B. halodurans* C-125 and compared it with the *B. subtilis* genome (Takami et al. 2000); this was the first report of an alkaliphilic *Bacillus* strain (Fig. 4.9). Whole-genome analysis gives us useful information about adaptation to alkaline environments.

Many alkaline enzymes genes were also detected, some of which have useful industrial applications. Two years later, another whole-genome sequence of *Oceanobacillus iheyensis* HTE831 was determined, which is an alkaliphilic and extremely halotolerant bacterium from a deep-sea sediment in the Mariana Trench.

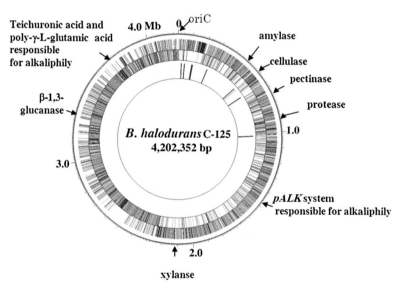

Circular structure of *Bacillus halodurans* C-125 gene

Fig. 4.9 Complete genome sequence of *Bacillus halodurans* (*B. halodurans*) C-125

The genome consists of 3.6 Mbp, encoding proteins such as those potentially associated with roles in regulation of intracellular osmotic pressure and pH homeostasis (Horikoshi 2006, 2011).

Alkaline Enzymes

Alkaline Proteases (Horikoshi 1999b)

In 1971, I reported the production of an extracellular alkaline serine protease from alkaliphilic *Bacillus clausii* 221 (Fig. 4.10). This strain, isolated from soil, produced large amounts of alkaline protease that differed from the subtilisin group. As shown in Fig. 4.11, the optimum pH of the purified enzyme was 11.5 with 75 % of the activity maintained at pH 13.0. The enzyme was completely inhibited by diisopropylphosphate or 6 M urea, but not by ethylenediaminetetraacetic acid or *p*-chloromercuribenzoate. The molecular weight of the enzyme was 30,000, which is slightly higher than that of other alkaline proteases. The addition of 5 mM calcium ions revealed a 70 % increase in activity at the optimum temperature (60 °C). Figure 4.12 shows crystals of alkaline protease no. 221 (Horikoshi 1971b).

Twenty years later, this enzyme gene was cloned and the sequence determined in 1992 (accession number S48754). We reported that the gene encoding an alkaline serine protease from alkaliphilic *Bacillus clausii* 221 was cloned in *Escherichia coli* and expressed in *Bacillus subtilis*. The deduced amino acid sequence of the alkaline protease from *Bacillus clausii* 221 had higher homology to the protease from other alkaliphilic bacilli.

Isolation of Detergents and H₂O₂-Resistant Alkaline Proteases

Alkaline proteases are used extensively in detergents, the food industry, and leather tanning. Enzymes produced commercially are derived only from microorganisms, and the microorganisms must be able to produce a high enzyme yield from low-cost substrates. The success of alkaline proteases in detergents is dependent on whether the enzymes have the following properties: (1) a wide pH activity range, (2) stability under high alkaline conditions, (3) high activity and stability in the presence of surfactants, (4) high stability in the presence of builders such as chelating reagents and bleaching agents, (5) high activity over a wide temperature range, (6) long shelf-life, and (7) low production cost. Although many enzymes have been reported, the alkaline proteases described here have several weak points in their enzymatic properties, such as being sensitive in the presence of oxidants and chelating agents. These disadvantages have been overcome by the isolation of new *Bacillus* strains by the author's group. Details are given next.

[Agr. Biol. Chem., Vol. 35, No. 9, p. 1407~1414, 1971]

Production of Alkaline Enzymes by Alkalophilic Microorganisms

Part I.   Alkaline Protease Produced by *Bacillus* No. 221

By Koki HORIKOSHI

The Institute of Physical and Chemical Research, Wako-shi, Saitama-ken, 351
Received March 10, 1971

A crystalline alkaline protease was prepared from *Bacillus* No. 221 isolated from soil. The characteristic point of this microorganism is especially good growth in alkaline media. The enzyme was most active at pH 11.5~12 towards casein and was stable at pH values from 4 to 11 on 10 min incubation at 60°C. Calcium ion was effective to stabilize the enzyme especially at higher temperatures. The enzyme was completely inactivated by DFP and urea, but not affected by sulfhydryl reagent, EDTA, SLS, and DBS. The specific activity of the enzyme towards casein was about 18,000 unit/mg. and the isoelectric point was higher than pH 9.4. The molecular weight and sedimentation constant was approximately 30,000 and 3.5 S respectively, and N-terminal of the enzyme was identified to be alanine. The results indicate that the No. 221 alkaline protease is different from those of alkaline proteases of *Bacillus subtilis*.

It has been reported that a certain species of bacteria can grow in high alkaline media containing high concentration of sodium carbonate or sodium bicarbonate.[1] Some of these bacteria produce a large amount of alkaline protease, and neither neutral nor acid protease can be detected in the culture broth.

Bacillus No. 221 isolated from soil accumulates a large amount of alkaline protease which is entirely different from *Bacillus subtilis* alkaline proteases.[2~4] The newly isolated enzyme of *Bacillus* No. 221 is most active at pH 11~12. This paper deals with isolation of the strain *Bacillus* No. 221 and some properties of the crystalline alkaline protease.

MATERIALS AND METHODS

Medium. An isolation medium (I-medium) contained: glucose, 10 g; polypeptone, 5 g; Difco yeast extract, 5 g; K_2HPO_4, 1 g; $MgSO_4$ $7 H_2O$, 0.2 g; Na_2CO_3, 10 g and 1 liter of distilled water. Sodium carbonate was sterilized separately and added to the medium. The I-medium was solidified by the addition of agar (1.5% w/v), if necessary.

Isolation method of alkalophilic bacteria. A small amount of soil was suspended in sterilized water and spread on I-medium agar plates. The plates were incubated at 37°C for 24 to 48 hr. *Bacillus* No. 221 producing large amount of alkaline protease was isolated from about 400 colonies.

Characterization and identification of bacteria. Microbiological properties were investigated according to the methods described in "Aerobic Sporeforming Bacteria"[5] and "Bergey's Manual of Determinative Bacteriology."[6] Unless stated otherwise, media used

1) K. Horikoshi, *Proc. Agr. Chem. Soc. Japan*, p. 201 (1971).
2) A. V. Guntelburg and M. Ottesen, *Compt. Rend. Trav. Lab. Carlsberg*, **29**, 36 (1954).
3) B. Hagihara, "The Enzyme," Vol. 4, ed. by Boyer, Lardy, and Myrback, Academic Press Inc., New York, 1958, p. 193.
4) D. Tsuru, H. Kira, T. Yamamoto and J. Fukumoto, *J. Agr. Biol. Chem.*, **30**, 1261 (1966).

5) N. R. Smith, R. E. Gordon and F. E. Clark, "Aerobic Sporeforming Bacteria," U.S. Dept. of Agric., 1952.
6) R. S. Breed, E. G. D. Murray and N. R. Smith, "Bergey's Manual of Determinative Bacteriology," Williams and Wilkins Co., Baltimore, 1957.

Fig. 4.10   The first report on the production of alkaline enzymes

Fig. 4.11 Alkaline protease having optimum pH at 11.5

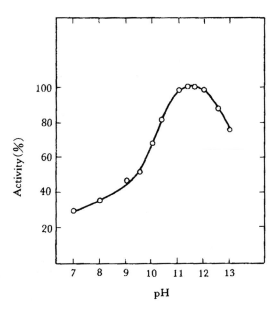

Fig. 4.12 Crystals of alkaline protease no. 221

E-1 Enzyme of *Bacillus cohnii* D-6

In 1972, we discovered a very stable alkaline protease in the presence of detergents containing high concentration of perborate in the absence of calcium ion. *Bacillus* sp. no. D-6 (FERM no. 1592; later designated *Bacillus cohnii* D-6) produced an alkaline protease, E-1, that was more stable in the presence of detergent additives at 60 °C than *Bacillus clausii* 221 protease.

The purified E-1 had a molecular mass of about 43 kDa as estimated by sodium dodecyl sulfate-polyacrylamide gel electrophoresis (SDS-PAGE) and a specific activity of 115 units/mg protein at pH 10.5 in 50 mM borate buffer. Isoelectrofocusing analysis indicated the pI value of the enzyme to be pH 9.7–9.9. The N-terminal amino acid sequence determined was Asn-Asp-Val-Ala-Arg-Gly-Ile-Val-Lys-Ala-Asp-Val-Ala-Gln. In the absence of calcium ions, caseinolytic activity of E-1 at 35 °C was observed in a wide pH range of 6–12, with optimum pH around 10–11 in 20 mM glycine-NaOH buffer. E-1 was stable at pH 6–12 but unstable below pH 5 and above pH 13 after a 24-h incubation at 25 °C. The optimal temperature for activity at pH 10 was 65 °C. The enzyme was stable up to 65 °C after a 30-min incubation at pH 10 in the presence of 5 mM $CaCl_2$. Phenylmethylsulfonyl fluoride and diisopropyl fluorophosphate (1 mM each) inhibited the enzyme activity by 98 and 92 %, respectively. Chelating agents, such as *O*-phenanthroline, EDTA, and EGTA (5 mM each), did not inhibit the enzyme activity at all.

The most striking property of the E-1 enzyme is its strong resistance to oxidants such as H_2O_2, as well as to chelating reagents such as EDTA and EGTA. If the enzyme productivity could be increased, the E-1 enzyme would be the best enzyme for detergent additives. After our discovery, many researchers tried to industrialize it, but without success. Almost 30 years later, the productivity my colleagues noted was dramatically increased to more than 10 g/l using gene technology.

Starch-Degrading Enzymes

No reports concerning optimal activity in the alkaline pH range were published before 1971 despite extensive study by many researchers. One noted Japanese enzymologist, Prof. Juichiro Fukumoto of Osaka City University, who investigated amylase, used to say, "There are no alkaline amylases in nature, because I spent over 30 years looking but was not able to discover any."

I attempted to isolate alkaliphilic microorganisms producing alkaline amylases, and in 1971 an alkaline amylase was produced in Horikoshi-II medium by cultivating alkaliphilic *Bacillus* sp. no. A-40-2 (Horikoshi 1971c). Several types of alkaline starch-degrading enzymes were subsequently discovered by cultivating alkaliphilic microorganisms.

α-Amylases

Bacillus pseudofirmus A-40-2 Amylase

Production of the alkaline amylase was first achieved in the alkaliphilic *Bacillus pseudofirmus* A-40-2 (A TCC21592), which was selected from about 300 colonies of bacteria grown in Horikoshi-II medium. The isolated strain was an aerobic, spore-forming, gram-positive, motile, rod-shaped bacterium with peritrichous flagella.

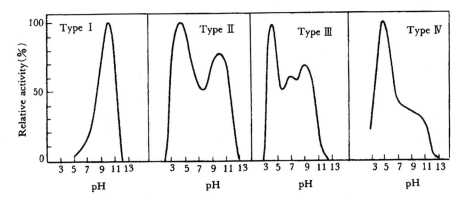

Fig. 4.13 Four types of alkaline amylases: Type I, α-amylases; Types II, III, and IV produce cyclodextrins

It was clear that the bacterium belonged to the genus *Bacillus*. Although the morphological, cultural, and biochemical characteristics of the strain resembled those of *Bacillus subtilis*, the special feature of the bacterium was that growth was very good in alkaline media, and the optimal pH for growth was about 10. No growth was detected in neutral media.

Bacillus pseudofirmus A-40-2 was grown aerobically at 37 °C in Horikoshi-II medium for 3 days. The alkaline amylase in the culture fluid was purified by a DEAE-cellulose column and a hydroxyl apatite column followed by gel filtration. The molecular weight was estimated to be about 70,000 by the gel filtration method. The enzyme is most active at pH 10.0–10.5 and retains 50 % of its activity between pH 9.0–11.5 (Fig. 4.13). The enzyme is not inhibited by 10 mM EDTA at 30 °C but is completely inactivated by 8 M urea. The enzyme can hydrolyse 70 % of starch to yield glucose, maltose, and maltotriose, and it is a type of saccharifying α-amylase.

Discovery and Properties of No. 38-2 and No. 17-1 Cyclodextrin Glycosyltransferases (CGTases)

We selected two strains, *Bacillus* sp. no. 38-2 and no. 17-1, as the best enzyme producers from starch-degrading alkaliphilic strains. These bacteria were aerobic, spore forming, motile, and rod shaped. All the strains belong to the genus *Bacillus*. The organisms were aerobically cultivated for 3 days in Horikoshi-II medium at 37 °C. The crude enzymes in the supernatant fluids were purified by starch adsorption followed by conventional DEAE cellulose column chromatography.

The crude enzyme in the culture broth of *Bacillus* sp. no. 38-2 was a mixture of three enzymes: acid CGTase, having optimal pH for enzyme action at 4.6; neutral CGTase, pH 7.0; and alkaline CGTase, pH 8.5. A Southern blot hybridization experiment showed that only one band hybridised with the *Bacillus* sp. no. 38-2 CGTase gene. The crude enzyme of *Bacillus* sp. no. 17-1 was a mixture of two enzymes: acid CGTase, optimal pH 4.5, and alkaline CGTase, optimal pH 9.5. Since our

Table 4.5 Properties of alkaline amylases

Type	Strain no.	Optimal pH	Ratio of activity	Stable pH	Protection by Ca^{2+}
I	A-40-2	10.5	1.0	7.0–9.5	+
	A-59	10.5	1.0	7.0–9.5	+
	27-1	10.5	1.0	7.0–9.0	+
	124-1	10.5	1.0	7.0–9.0	+
II	135	4.0–4.5; 10	0.8	7.0–9.0	+
	169	4.0–4.5; 10	0.7	7.0–9.0	+
III	38-2	4.5; 9.0	0.6	5.0–10.5	+
IV	13	4.5	0.3	6.5–10	+
	17-1	4.5	0.35	6.5–10	+

Fig. 4.14 Molecular model of β-cyclodextrin

discovery of the bacterium alkaliphilic *Bacillus* sp. no. 38-2, many alkaliphilic microorganisms producing CGTases have been reported (Table 4.5).

Figure 4.14 shows the molecular model of β-cyclodextrin, and Fig. 4.15 describes an industrial production process of β-cyclodextrin. As shown in Fig. 4.16 and Table 4.6, many industrial applications of cyclodextrins have been reported (Horikoshi 2006).

Cellulases of Alkaliphilic Bacillus Strains

Cellulases for Cellulose Waste (Horikoshi 1999a)

About three decades ago, the percentage of flush toilets in Japan was very low. Japanese microbiologists sought alkaline cellulase that hydrolysed native cellulose to treat human excrement. During the conventional treatment in septic tanks,

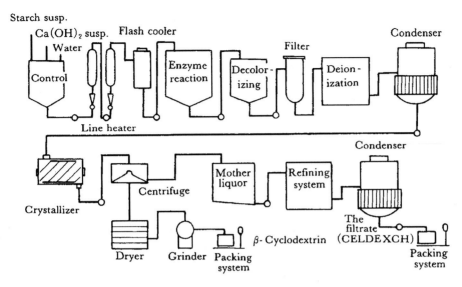

Fig. 4.15 An industrial production process of β-cyclodextrin. (*susp.* suspension)

Fig. 4.16 Wasabi paste for "sushi" made by inclusion of compounds of flavour and taste

the pH value used to be between 8 and 9 because of the ammonia produced. Therefore, no cellulolytic enzyme so far reported could hydrolyse the cellulose in human excrement.

Table 4.6 Industrial applications of cyclodextrins

Functions	Guests	End products
Foods		
1. Emulsification	Oils and fats	Margarine, cake, whipping cream, French dressing
2. Stabilization	Flavours, spices, colours, and pigments	Horseradish paste, mustard paste, cakes and cookies, pickled vegetables, dried vegetables
3. Masking of taste and odour		Juices, soy milk, bone powder, boiled rice
4. Improvement of quality		Hard candy, cheese, soy sauce, canned citrus fruits and juices
5. Reduce volatility	Ethanol	Food preservatives
6. Others		Breath mints
Cosmetics and toiletries		
1. Emulsification	Oils and fats	Face cream, face lotion, toothpaste
2. Stabilization	Flavours and fragrances	Bath refresher crystals
Agrochemicals		
1. Stabilization	Pyrolnitrin	Fungicide
	Pyrethroids	Insecticide
2. Reduce volatility	Organic phosphates (DDVP)	Insecticide
	Thiocarbamic acid	Herbicide
3. Reduce toxicity	2-Amino-4-methyl-phosphynobutyric acid	Fungicide

Functions	Guests and end products
Pharmaceuticals	
1. Improve solubility	Prostaglandins, steroids, cardiac glycosides, nonsteroidal antiinflammatory agents, barbiturates, phenytoin, sulfonamides, sulfonylureas, benzodiazepines
2. Chemical stabilization	
A. Hydrolysis	Prostacyclin, cardiac glycosides, aspirin, atropine, procaine
B. Oxidation	Aldehydes, epinephrine, phenothiazines
C. Photolysis	Phenothiazines, ubiquinones, vitamins
D. Dehydration	Prostaglandin E_1, ONO-802
3. Improve bioavailability	Aspirin, phenytoin, digoxine, acetohexamide, barbiturates, nonsteroidal antiinflammatories
4. Powdering	ONO-802, clofibrate, benzaldehyde, nitroglycerin, vitamin K_1, K_2, methylsalicylate
5. Reduce volatility	Iodine, naphthalene, *d*-camphor, *l*-menthol, methylcinnamate
6. Improve taste, smell	Prostaglandins, alkylparabens
7. Reduce irritation to stomach	Nonsteroidal antiinflammatory agents
8. Reduce hemolysis	Phenothiazines, flufenamic acid, benzylalcohol, antibiotics

Isolation of No. 212 Enzyme

In 2 years of screening, we isolated alkaliphilic *Aeromonas* sp. no. 212 (ATCC31085), producing a cellulase that hydrolysed cellulose in human faeces. The culture medium is 0.5 % ammonium sulfate, 1.5 % pulp flock, 0.02 % glucose, 0.1 % yeast extract, 0.02 % $MgSO_4$ $7H_2O$, and 0.2 % K_2HPO_4 containing 1 % $NaHCO_3$. The strain was cultured at 37 °C for 3 days.

The enzyme complex is a mixture of at least six enzymes. The crude preparation has the following properties: optimum pH for hydrolysing activity is 5–9, and it is thermostable up to 60 °C. The enzyme preparation acts on cellulose, microcrystalline cellulose, filter paper, swollen cellulose, absorbent cotton, and carboxylmethylcellulose.

Degradation of Faecal Cellulose by No. 212 Enzyme

Small-scale test: A quantity of 15 g toilet paper, microcrystalline cellulose, and cellulose powder of faecal cellulose in 1 l liquid (pH 8.0) was tested as a substrate. The weight loss of the substrate solid was measured.

Bench-scale test: A quantity of 9.5 l excrement-digested sludge was poured into each of four brown glass excrement digestion tanks (diameter 20 cm, height 32 cm, effective volume 10 l), with temperature maintained at 37 °C. The change in the amount of accumulated scum was observed with the naked eye. The thickness of scum in the cellulase-added tank was about one-third that of the control. Cellulase of no. 212 decomposed the cellulose in human excrement quite well and reduced the thickness of the scum.

The Japanese economy, however, slowly began to grow. Flush toilets became popular and the classic toilet disappeared, so no plant was ever constructed.

Alkaline Cellulase from Alkaliphiles

No enzyme with an alkaline optimum pH for activity (pH 10 or higher) had been reported before our publication (Horikoshi 2006). We isolated many alkaliphilic bacteria producing alkaline cellulases. One of these, alkaliphilic *B. cellulosilyticus* N-4, produced at least three CMCases. Another bacterium, *B. akibai* 1139, produced one CMCase that was entirely purified, and the enzyme had an optimum pH for activity at pH 9.0. All alkaline cellulases so far discovered could not hydrolyse or could only slightly hydrolyse cellulose fibers. Our finding of alkaline cellulases paved the way for the industrial application of these enzymes as a laundry detergent additive, and hundreds of scientific papers and patents have been published.

I found bacterial isolates (*Bacillus* no. N4 and no. 1139) producing extracellular alkaline carboxymethylcellulases (CMCases). One of these, alkaliphilic *Bacillus cellulosilyticus* N-4, produced multiple CMCases that were active over a broad pH range (pH 5–10) (Fig. 4.17). Several cellulase-producing clones that have different DNA sequences were obtained. Another bacterium, *Bacillus akibai* no. 1139,

Fig. 4.17 Activity and pH
curve of alkaline cellulase
N-4

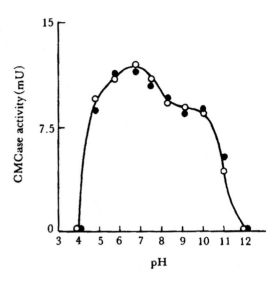

Fig. 4.18 Alkaline cellulase
having optimal pH at 9

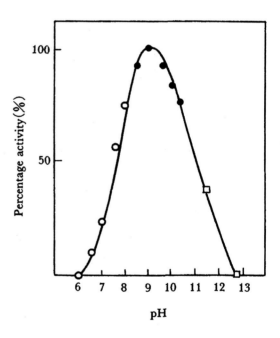

produced one CMCase, which was purified and shown to have optimum pH for
activity at pH 9.0 (Fig. 4.18). The enzyme was stable over the range pH 6–11 (24 h
at 4 °C and up to 40 °C for 10 min). The CMCase gene of *Bacillus akibai* 1139 was
also cloned in *E. coli*. We constructed many chimeric cellulases from *B. subtilis* and
B. cellulosilyticus N-4 enzyme genes to understand the alkaliphily of N-4 enzymes.

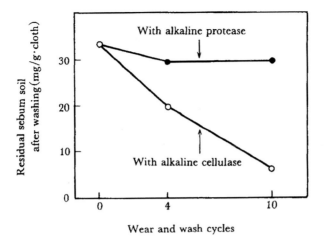

Fig. 4.19 Detergent effect of alkaline cellulase

Cellulases as Laundry Detergent Additives

The discovery of alkaline cellulases created a new industrial application of cellulase as a laundry detergent additive. Susumu Ito of Kao Company in Japan mixed alkaline cellulases with laundry detergents and investigated the washing effect by washing cotton underwear. The best results were obtained by one of the alkaline cellulases produced by an alkaliphilic *Bacillus* strain. Some cellulases do not attack cotton but release dirt attached to amorphous cellulose fibers (Fig. 4.19). Fabrics are mostly crystalline cellulose, and this kind of cellulose is never degraded by our alkaline cellulases. Several laundry detergents containing the alkaline cellulase are now commercially available. In Japan, such laundry detergents enjoy approximately 80 % of the market. To the best of my knowledge, this is the first-large scale industrial application of cellulase (Horikoshi 2011). I am convinced that no laundry detergent containing alkaline cellulase would have been developed if we had not found and reported the production of alkaline cellulase by alkaliphilic bacteria (Fig. 4.20).

Other Enzymes (Horikoshi 1999a, 2006)

Xylanases

The first report of xylanase of alkaliphilic bacteria was published in 1973 by Horikoshi and Atsukawa. The purified enzyme of *Bacillus* sp. no. C-59-2 exhibited a broad optimal pH ranging from 6.0 to 8.0. Then, Okazaki reported that four thermophilic alkaliphilic *Bacillus* stains (Wl, W2, W3, W4) produced xylanases (Okazaki et al. 1984, 1985). The pH optimum for enzyme action of strains Wl and W3 was 6.0 and that for strains W2 and W4 was between 6 and 7. The enzymes

Fig. 4.20 New laundry detergent (*right, small box* "Attack") containing alkaline cellulase

were stable between pH 4.5 and 10.5 at 45 °C for 1 h. The optimal temperatures of xylanases of W1 and W3 were 65 °C and those of W2 and W4 were 70 °C. The degree of hydrolysis of xylan was about 70 % after 24-h incubation.

Subsequently, two xylanases were found in the culture broth of *Bacillus halodurans* C-125 (Honda et al. 1985). Xylanase A had a molecular weight of 43,000 and that of xylanase N was 16,000. Xylanase N was most active at pH 6–7; xylanase A was most active at a pH range of 6–10 and had some activity at pH 12. The xylanase A gene was cloned, sequenced, and expressed in *E. coli*.

After the demonstration that alkali-treated wood pulp could be biologically bleached by xylanases instead of by the usual environmentally damaging chemical process using chlorine, the search for thermostable alkaline xylanases has been extensive.

Mannan-Degrading Enzymes

β-Mannan is a kind of hemicellulose contained in higher plants such as konjac, guargum, locust bean, and copra, and it easily dissolves in alkaline water. Mannan-degrading enzymes of neutrophilic bacteria, actinomyces, and fungi have been studied. However, no mannan-degrading enzyme that hydrolyzes under alkaline conditions had been reported before our discovery. We reported the isolation and properties of mannan-degrading microorganisms.

A small amount of soil was spread on agar plates containing 1 % β-mannan from larch wood, 1 % polypeptone, 0.2 % yeast extract, 0.1 % KH_2PO_4, 0.02 % $MgSO_4$ $7H_2O$, and 0.5 % sodium carbonate. The plates were incubated at 37 °C for 48–72 h. Strain AM-001 with a large clear zone around the colony was selected as the enzyme producer. The isolate grew at temperatures from 20 to 45 °C, with an optimum at

37 °C in the medium described above. The pH range for growth was 7.5–11.5 with the optimum at pH 8.5–9.5. The taxonomic characteristics of this alkaliphilic *Bacillus* strain were almost the same as those of *Bacillus hortii*.

The enzyme produced in culture broth contained three β-mannanases and β-mannosidase. Two β-mannanases were most active at pH 9.0, and one demonstrated optimum enzyme action at pH 8.5. These properties of the enzyme are good for the production of D-mannose from β-mannan in the presence of the β-mannanases described here. Since then, several mannan-degrading alkaliphiles have been published, although no industrial application has been established yet.

Pectin-Degrading Enzymes

Bacillus sp. No. P-4-N Polygalacturonase

The first paper on alkaline endo-polygalacturonase produced by alkaliphilic *Bacillus* sp. no. P-4-N was published in 1972. *Bacillus* sp. no. P-4-N could grow in the pH range from 7.0 to 11.0, but the most active growth was observed in a pH 10 medium containing 1 % Na_2CO_3. The medium of pH 10.4 contained 1 % Na_2CO_3, 0.005 % $Mn_2SO_4 \cdot 7H_2O$, and 3 % pectin as essentials. The enzyme was most stable at pH 6.5 and up to 70 °C in the presence of Ca^{2+}, but was inactivated completely at 80–90 °C.

Recently, the author's group reidentified this strain by 16 S RNA analysis, and *Bacillus* sp. P-4-N was found to be very similar to or the same strain as *Bacillus halodurans* C-125.

Industrial Applications of Alkaline Pectinase: Production of Japanese Paper (*Washi*)

Alkaliphilic bacteria were isolated form soil, sewage, and decomposed manure in Japan and Thailand. *Bacillus* sp. no. GIR 277 had strong macerating activity towards Mitsumata bast. The bacteria isolated were motile, aerobic, spore-forming rods and grew well on nutrient agar of pH 9.5 adjusted with Na_2CO_3. This bacterium produced pectate lyase, which had an optimal pH for enzyme action at 9.5. Japanese paper was produced as follows. Four grams of Mitsumata bast was suspended in 100 ml culture medium containing 0.05 % yeast extract, 0.05 % casamino acids, 0.2 % NH_4Cl, 0.1 % K_2HPO_4, and 0.05 % $MgSO_4 \cdot 7H_2O$. After sterilization, Na_2CO_3 was added to a concentration of 1.5 %, and *Bacillus* sp. no. GIR 277 was inoculated into the medium. After 5 days of cultivation at 30 °C with shaking, the retted bast was harvested and Japanese paper was prepared by the method described in Japanese Industrial Standard P8209. The overall yield of pulp was about 70 %. The strength of the unbeaten pulp resulting from bacterial retting was higher than that obtained by the conventional soda ash-cooking method. The paper sheets were very uniform and soft to the touch. Yoshihara and Kobayashi also concluded that bacterial retting under alkaline condition is a potentially useful process for the production of pulp of excellent quality from non-woody pectocellulosic fibres.

Lipases

Alkaline lipases have been isolated from many microorganisms, as described in the previous book written by the present author. However, few alkaliphilic *Bacillus* strains have been reported as alkaline lipase producers. Mostly, non-alkaliphilic *Bacillus* such as *Bacillus stearothermophilus* have mainly been isolated and their lipases extensively reported.

A Japanese company has produced a laundry detergent containing a fungal lipase produced by recombinant DNA technology. The fungus itself is not alkaliphilic, but the enzyme is alkaline lipase. This is the only industrial application of lipase as a laundry detergent additive.

Future of Alkaliphiles

Since the discovery of alkaliphilic bacteria, more than 5,000 papers on many aspects of alkaliphiles and alkaliphily have been published. The alkaliphiles are unique microorganisms, with great potential for microbiology and biotechnological exploitation.

The aspects that have received the most attention in recent years include (1) extracellular enzymes and their genetic analysis, (2) mechanisms of membrane transport and pH regulation, and (3) the taxonomy of alkaliphilic microorganisms. What will be the next line of development? It is unclear at present, but it may be the wider application of enzymes (Fig. 4.21).

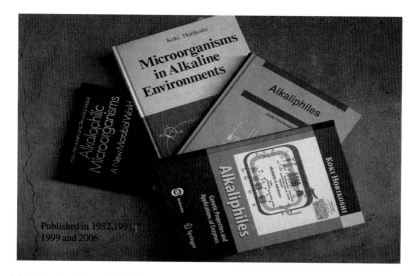

Fig. 4.21   Four books on alkaliphiles

Chapter 5
Ingham Family and Biosciences

Inspection Tour: University and Research Institutions in UK

In October 1982 I received an invitation from the British Embassy in Tokyo to visit universities and research institutions in the United Kingdom, with all travel expenses paid by the Embassy. I accepted the invitation, of course. We were to visit Oxford University, Cambridge University, Imperial Chemical Industries, Unilever PLC, and Rothamsted Experimental Station. While in Britain I encountered several scientists who were to have a great influence on my research.

Joel Mandelstam

I visited Professor Joel Mandelstam's laboratory at Oxford University on October 22. The professor was working on X-ray crystallography and he showed me an X-ray analyser that he had made himself and used to determine the structures of several proteins. I asked him, if I brought him my alkaline enzymes, could he determine their structures? He replied that of course he could, but he strongly suggested that I buy the same X-ray analyser in Japan. And he looked at me with a grin. Yes, it was true: if I used a commercially available instrument, I could not claim the credit for being the first to conduct the experiment. Even now I still remember his homemade apparatus.

Bill Grant

We visited Professor Bill Grant at Leicester University to discuss spores of bacteria. I had met him by chance in 1980 at Cambridge railway station and talked with him for half an hour. He said that he would like to visit my laboratory in Japan.

© Springer Japan 2016
K. Horikoshi, *Extremophiles*, DOI 10.1007/978-4-431-55408-0_5

In 1984 the ERATO Superbugs project started, and he joined us in the project and discovered triangular archaea.

Sydney Brenner

At the Medical Research Council (MRC) in Cambridge, Professor Brenner was focused on establishing *Caenorhabditis elegans* as a model organism for the investigation of animal development, including neural development. On October 27, he told me: "Frederick Sanger is working on DNA sequencing, although he does not talk about it. The only thing I can do is just wait for his successes."

Sydney came to RIKEN the following year and he explained his wonderful research work on *Caenorhabditis elegans*. He also received the 2002 Nobel Prize in Physiology or Medicine.

Rothamsted Experimental Station

The scientists at the station were improving the yield of green peas. They made green peas with short tendrils, slender stems, and small leaves. They thought that if the energy used to make these parts of the plant could be reduced, then the energy saved would be transferred to the peas themselves. However, the photo-energy derived from sunlight was insufficient because of the small leaves. As the result, the yield of green peas fell. I was really surprised at their way of thinking, which was so different from the Japanese way.

During the 10-day tour (October 18–28), I had a splendid interpreter in Kazuko Yoshida. I retain the happiest of memories of my tour of England.

Ingham Family and Our Family

Before Deep-Sea Research

In 1983, I had a telephone call from Kazuko. She said: "I and my husband Peter Ingham are in Kanazawa. We'll be here until 1986." He had a post as a visiting lecturer at Kanazawa University, College of Liberal Arts.

Sachiko and I met with the Inghams in Roppongi, Tokyo. We talked about all sorts of things for hours. Since then, my wife and I have become firm friends with Kazuko and Peter, a journalist on *The Times*. Even today I continue to encounter new ways of thinking (Fig. 5.1).

Fig. 5.1 *From the left,*
Peter, Ellen (Peter's mother),
and Kazuko

Sachiko's father Shigeo had lived in Chiswick, West London, in 1921. Peter and Kazuko also live in Chiswick. I developed a fondness for Chiswick and I decided to buy a property in the area. Kazuko kindly helped me to research the possibilities and finally I bought a flat there.

In 1991, I received an invitation from the International Institute of Biotechnology. I was to be awarded the Institute's Gold Medal and I was expected to give a lecture. The first thing I did was telephone Peter to make sure that my reply would be polite, elegant, and in accordance with the protocol, and asked him to edit and improve my draft speech.

Biosciences in JAMSTEC

In 1993 nobody was working on biology in JAMSTEC so I devised methods to achieve our ambitions. I consulted Peter on a name for the research project. He suggested "Deep Star" (*D*eep *S*ea *E*nvironment *E*xploration *P*rogramme, *S*uboceanic *T*errain *A*nimalcule *R*etrieval), which is still used in this field.

On February 28, 1996, the submersible *Kaiko* descended to 11,000 m beneath the sea and sent back photographs of the seabed (see Chap. 9).

I told Peter, then working as a journalist on *The Times*, about this, and on March 18 *The Times* carried a large article on the event. However, there was not a single mention in the Japanese newspapers.

When I returned to Tokyo, I received an unexpected telephone call from Mr. Meguro, a chamberlain of the Imperial Household Agency, saying that the Emperor had read the article in *The Times* and would like to hear more about our project.

I had an audience with the Emperor on July 2 at the palace and explained the details of the project (see Chap. 10). This was the turning point for research into deep-sea biology in Japan because the government, suddenly realising the importance of the event, woke up to the field and start funding research programmes. On March 28, 2001, the Emperor and Empress visited JAMSTEC to observe our deep-sea organisms

On July 10 I was awarded a doctor's degree at Canterbury Cathedral and delivered my thoughts in a lecture which Peter and Professor Alan Bull, of the University of Kent, had kindly edited.

It is not an overstatement to say that, thanks to Peter, the door was opened to Japanese research into this new field.

Chapter 6
Superbugs Project

In the middle of 1984, Dr. Gen-ya Chiba, vice-president of the ERATO Project, visited my office at RIKEN. He told me that he was going to set up a fourth ERATO project and that I was a strong candidate for what was called the abnormal microbes research project. I asked what the definition of abnormal microbes was and how the research was to be funded. To my surprise he told me that the definition of the microbes was up to me and that the funding would be 1 million dollars a year for 5 years.

I could invite biologists from all over the world if I wanted. Furthermore, Dr. Chiba said: "You would have complete freedom in the research."

Needless to say, my answer was an enthusiastic yes.

ERATO Superbugs Project

The 5-year Superbugs Project of the Exploratory Research for Advanced Technology (ERATO) programme in the Japan Research and Development Corporation (JRDC) was launched in 1984 to search the world's more extreme environments for microorganisms that grow in conditions of strong pH or conditions of extreme temperature, salinity, or pressure, and to try to use their unique properties to establish a "new biotechnology." I was in charge of the project, which was initially called "the project for the study of microorganisms in extreme environments." This was too cumbersome to say, so I renamed it the "Superbugs Project."

What the Superbugs Project aimed to do was to study the cell structure and function, metabolism, and enzyme actions of microorganisms in extreme environments. The information obtained during the course of the project would, we hoped, make an important contribution to modern biotechnology. To study superbugs, the following three groups were set up in the Tokyo metropolitan area (Figs. 6.1 and 6.2).

© Springer Japan 2016
K. Horikoshi, *Extremophiles*, DOI 10.1007/978-4-431-55408-0_6

Fig. 6.1 Research building
for ERATO Research at
Komagome, Tokyo

Fig. 6.2 ERATO provided a mobile research bus with modern equipment for screening
extremophiles

Fundamental Work Group

Superbugs were isolated from all over the world during the course of the research period. The microbiology, enzymology, and molecular biology of those microorganisms were systematically investigated from the industrial point of view as well as for academic interest. An important question was why superbugs required such extreme environments. This is a simple question to ask but a very difficult one to answer. To solve this problem, the DNA, RNA, proteins, lipids, and polysaccharides in superbugs were analysed using modern technology. The information obtained by this group was to aid the other two research groups.

Metabolism and Production Group

Metabolic pathways in superbugs differ from those of "moderate" microorganisms. In particular, secondary metabolites found in superbugs were one of the targets because no microbiologist had yet studied them. Some halophilic bacteria have entirely different gene expression systems that are not detected in more "moderate" bacteria, and the expression and control mechanisms of tolerant enzymes were to be extensively studied and their potential abilities developed.

Tolerance Introduction Group

The introduction of some tolerance activities such as alkaliphily, halophily, or thermophily into "moderate" bacteria was one of the most interesting and fascinating projects. Enzymes that are stable at higher temperatures would be commercially produced by using "moderate" bacteria such as *Escherichia coli* or *Bacillus subtilis* having some unique "superbug" properties. Once these methods had been established, industries would be able to make new bioreactors with immobilised cells or enzymes that could operate at higher temperatures and in higher salt concentrations. These results will promote the further progress of modem biotechnology.

Some of the research results were as follows.

A Novel Triangular Archaeon *Haloarcula japonica*

Halophilic archaea sometimes exhibit unusual morphologies in liquid culture. Some of the cells in a culture may be ribbon shaped or disc shaped, and occasionally triangular or square forms have been observed. However, high proportions of triangles are not usually seen in cultures of haloarchaea. In the course of screening a large number of hypersaline sites for new isolates of haloarchaea, we isolated a haloarchaeon that, unusually, produced predominantly triangular cells in liquid culture (Horikoshi and Grant 1998; Horikoshi 2011).

Isolation of a Triangular Archaeon

Professor William Grant and my colleagues collected about 600 samples from various areas of high NaCl concentration in nature. Small amounts of material were suspended in CM medium containing yeast extract 10 g, casamino acid 7.5 g, trisodium citrate 3.0 g, KCl 2.0 g, $MgSO_4 \cdot 7H_2O$ 20.0 g, $MnCl_2 \cdot 2H_2O$ 0.36 mg $FeSO_4 \cdot 7H_2O$ 50 mg, and NaC1 200 g, in 1,000 ml water, pH 7.2–7.8. For the preparation of solid media, agar at 1.8 % final concentration was added. After 6-day cultivation in the liquid medium just described at 37 °C with continuous shaking, the cell shape was examined using a phase-contrast microscope.

All at once, Bill called out to me: "Koki, come and look down our microscope. Triangular bacteria are moving!" An isolate from a Japanese saltern soil located on the Noto Peninsula exhibited triangular cells in the liquid medium (Fig. 6.3). The organism (TR-1) is motile by flagella, non-spore forming, and has red flat cells that are predominantly triangular in shape, although rhomboidal cells are also observed. The cells are typically 2–5×0.3–0.5 μm in size in liquid media. A typical electron micrograph of the TR-l is shown in Fig. 6.4.

Fig. 6.3 (**a**) Triangular *Haloarcula japonica* in CM medium. (**b**) Circular *Haloarcula japonica* in CM medium containing a low concentration of Mg ion

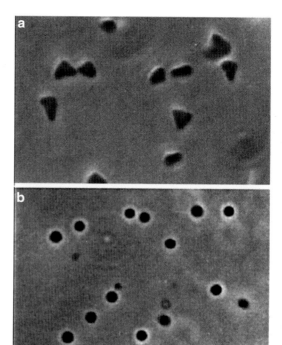

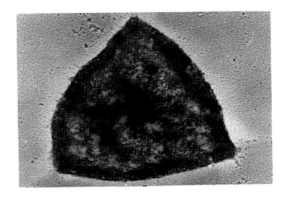

Fig. 6.4 Electron micrograph of *Haloarcula japonica*

Cell Division of Haloarcula japonica

The method of cell division of *Haloarcula japonica* attracted our scientific curiosity because of its characteristic shape. We analysed the course of cell division with time-lapse microscopic cinematography using a 16-mm movie camera at 15-s intervals in a plastic chamber at 42 °C. Figure 6.5 also shows photographs of the cell division processed by a colour image analyser. The average time for one cell cycle was 3.7 h.

Discovery of Organic Solvent-Tolerant Microorganisms

The discovery of new microbes is essential for the development of novel fields of microorganism research. In general, organic synthesis reactions differ from biochemical reactions in a reactive environment because the former occur in organic solvent solutions. On the other hand, biochemical reactions occur chiefly in aqueous solutions. Many biochemical reactions are inhibited by organic solvents. The scope of biochemical reactions would therefore be extended if they became possible in organic solvents. Biochemical reactions can occur in organic solvents if specific microbes develop tolerance to them. Organic solvent-tolerant bacteria are thought to be useful for investigating biochemical reactions in organic solvents. New microbial reactions may be found with the discovery of novel microorganisms.

Solvent-tolerant microorganisms also have numerous potential commercial applications in industrial biotransformation processes that involve the use of organic substrates with low solubility in water. When such compounds are used as substrates, large quantities of water are required, and water consumption is a major cost factor in the fermentation industry. The development of solvent-tolerant microorganisms or microbial catalysts for use in bioreactors might solve this problem. Therefore, a microorganism that can grow in organic solvents is needed.

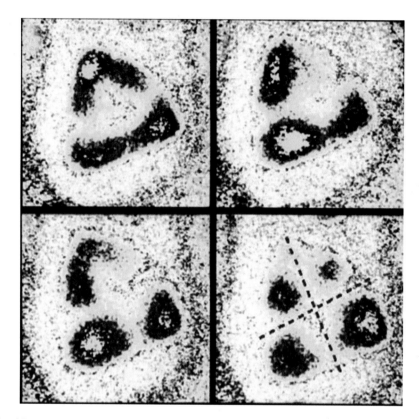

Fig. 6.5 Cell division of *Haloarcula japonica*

Toluene-Tolerant Bacteria

Toluene is a highly toxic solvent that kills most microorganisms at a concentration of 0.1 %. It has therefore been used for many years to sterilise microbial cultures and maintain solutions in a sterile condition. Furthermore, because toluene is effective in extracting lipids from cell membranes, it has been used as an unmasking agent in the assay of various enzymes.

From 800 soil samples collected from oil plants, oil wells, sludge, and tar sands in Japan and other countries, we isolated a toluene-tolerant bacterium that grew well in a nutrient medium containing toluene. To isolate toluene-tolerant microorganisms, a small amount of soil was suspended in sterile water, and 0.2 ml of this suspension was transferred to test tubes containing 5 ml bouillon medium. Toluene was then added to a final concentration of 30 % (v/v), the test tubes were plugged with butyl-rubber, and the cultures were incubated at 37 °C for 1 week in a test-tube shaker. As a result of screening, one bacterial strain that grew well in the toluene-containing medium was isolated from a soil sample collected in Kyushu, southern Japan. Based on morphological and biochemical characteristics of the strain determined using the standard methods, the isolate was designated *Pseudomonas putida* IH-2000.

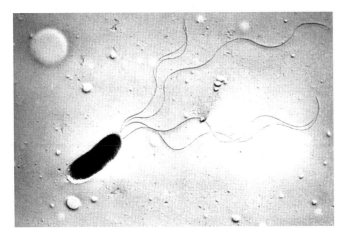

Fig. 6.6 Toluene-tolerant *Pseudomonas putida* IH-2000

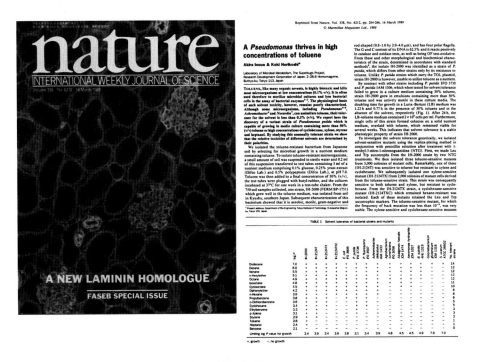

Fig. 6.7 Paper on *Pseudomonas putida* IH-2000 in *Nature*

We submitted a paper describing the toluene-tolerant *P. putida* IH-2000 to the journal *Nature* (Fig. 6.6), and the referee was initially sceptical. After we had answered numerous questions from the referee and provided a detailed account, the paper was finally accepted for publication (Fig. 6.7) (Inoue and Horikoshi 1989, 1991; Horikoshi 2011).

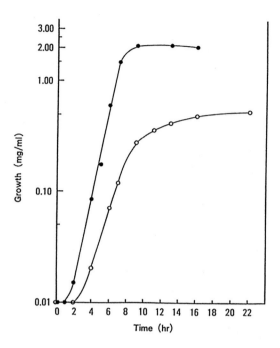

Fig. 6.8 Growth of *Pseudomonas putida* IH-2000. *Closed circles,* in absence of toluene; *open circles,* in the presence of toluene

Figure 6.8 shows the growth curve of strain IH-2000 in the modified LB medium with and without solvent. The doubling time for growth was 1.22 and 0.77 h in the presence of 30 % toluene and in the absence of the solvent, respectively. After 24 h, the LB medium with toluene added contained 2×10^9 cells/ml.

Interestingly, this strain was not tolerant to benzene, fluorobenzene, nitrobenzene, propanol, butanol, diethylether, propyleneoxide, chloroform, or ethylacetate. This strain could not grow in media containing 50 % ketone solvent, but exhibited growth at acetone and cyclohexanone concentrations of 2.5 %. The isolated strain is also resistant to styrene, *p*-xylene, ethylbenzene, cyclohexane, *o*-dichlorobenzene, propylbenzene, and various alkanes up to isooctane; thus, there is wide scope for producing both solvent-tolerant and solvent-degrading strains (Fig. 6.9). The marriage of these two desirable traits may be less than harmonious, however. The alterations in the cells of the toluene-tolerant strain that make growth in solvents possible could shield the cell from these solvents so perfectly that little penetration to the locations of degradative enzymes would be possible. We began a genetic analysis of the solvent resistance of *P. putida*, and isolated a series of mutants of new strains that are not resistant to toluene, but retain resistance to solvents of lower polarity in the order toluene > *p*-xylene > cyclohexane > hexane. By examining the partitioning index of these and other substances in a standard octanol/water mixture, we can predict accurately the resistance of these mutant strains to other compounds in the polarity scale.

In addition to mutant strains of *P. putida*, we can order other microorganisms in a similar hierarchy of solvent tolerance in which growth in a solvent of a given polarity indicates tolerance to solvents of lower polarity. Genetic analysis combined with biochemical and morphological studies provides useful insights into the nature

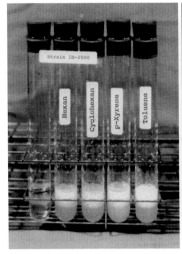

Fig. 6.9 Growth of *Pseudomonas putida* IH-2000 in organic solvents

of solvent tolerance in these bacteria. The outer membrane of gram-negative bacteria is important in determining the penetration of hydrophobic molecules through cell membranes, and consequently in determining the resistance of these bacteria to dyes, detergents, bile salts, and hydrophobic antibiotics. Mutations in the structure of the lipopolysaccharide of the outer membrane can profoundly affect permeability of these molecules, but these effects are complicated by concomitant changes in the amounts of specialised pore-forming, outer membrane proteins. Studies of the outer membranes of these solvent-resistant strains of *P. putida* has been rewarding.

Industrial Applications

Recently, my successor, Professor Noriyuki Doukyu, has extensively studied the mechanisms of adaptation and tolerance towards organic solvents, particularly in *Pseudomonas putida* and *Escherichia coli* strains (Fig. 6.10).

He has also worked on applications of organic solvent-tolerant microorganisms, such as bioconversion of cholesterol, bioconversion of indole to indigo, bioproduction of phenol from glucose, and bioremediation (Doukyu et al. 2003).

Isolation of Toxic Material-Insensitive Microorganisms

I would have liked to isolate microorganisms that could accumulate uranium in their cells. In fact I was able to find a few such bacteria, but further experimentation was strictly prohibited. My work ran counter to the Treaty on the Non-Proliferation of Nuclear Weapons, and all the microbes and radioactive isotopes were confiscated.

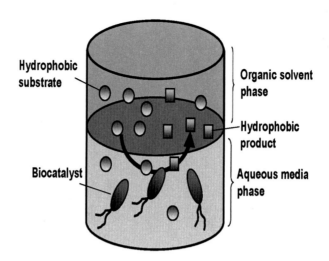

Fig. 6.10 Application of solvent-tolerant microorganisms

Instead, we isolated many bacteria insensitive to toxic metals, such as selenium, from a number of places.

From all over the world we isolated many methane-forming bacteria under extreme conditions, such as low or high temperatures, high alkaline pH values, or high concentration of salts.

Future of Extremophiles

Future extremophile research will progress with the isolation of further extremophiles using our new technology and will concentrate on the following questions:

1. Can we give tolerances to conventional microbes?
2. Can we find new materials for industrial applications?
3. Can we use extremophile DNA as a gene resource?
4. Can we improve conventional proteins using our gene resources?

I was unable to cultivate microbes under high pressure because of the limits of our technology. Today, however, we can cultivate piezophiles at pressures higher than 1,000 atm.

As yet, nobody can be sure whether our superbugs project has been a commercial success. Time will tell.

Chapter 7
Medal Lecture at the Royal Society

Appointment as Professor at the Tokyo Institute of Technology

In October 1987 I received an unexpected telephone call from the Tokyo Institute of Technology. "We would like to invite you to join the institute as a full professor because we are establishing a new department of biology. You are our choice of professor and we would be very happy if you would accept the invitation and set up a new academic unit for extremophiles. The department will open at the beginning of April 1988." I asked for a month to make up my mind.

The Tokyo Institute of Technology is the best university in Japan in the field of technology, but it had no biology department. This meant I had to start from scratch.

Of course, I talked it over with Sachiko. She said that I should move to the university and groom a successor. That way my name would last as the discoverer of alkaliphiles.

I then asked Professor Kin-ichiro Sakaguchi for his opinion. His answer was the same: go to the Tokyo Institute of Technology and groom a successor.

I discussed the matter with the president of RIKEN, who agreed with my decision. He asked that I should hold two posts (at RIKEN and the Tokyo Institute of Technology). I phoned the university and agreed to an academic unit consisting of a full professor, an associate professor, and two assistant professors.

In 1988, we had two graduate students: Masahiro Ito (Professor of Toyo University) and Hiroyasu Ogino (Professor of Osaka City University). They are now working on extremophiles in their laboratories. In June 1990, brand-new buildings were constructed in Nagatsuda, Yokohama.

Letter from the International Institute of Biotechnology

On 20 October 1990, an airmail letter from the International Institute of Biotechnology (IIB) in London landed on my desk at the Tokyo Institute of Technology. The letter said that Prince Michael of Kent would like to present me with the gold medal of the

© Springer Japan 2016
K. Horikoshi, *Extremophiles*, DOI 10.1007/978-4-431-55408-0_7

IIB at The Royal Society on 11 March 1991. At the presentation ceremony I would be expected to give an address on extremophile research (Fig. 7.1). Naturally, my answer was yes, but I was unsure of the correct form of reply. I phoned Peter Ingham (the husband of Kazuko Yoshida) and asked him to compose a suitably phrased letter in reply.

There were many things to think about. The lecture had to be understandable to the audience. How I could give a lecture about my work in pure English rather than American English?

Usually when giving a lecture, I begin: "Mr. Chairman, ladies and gentleman." But what should I say on this occasion? Finally I settled on: "Your Royal Highness, Prince Michael of Kent." Should I shake his hand? Nobody knew, so I thought, just go for it.

Now, the next thing I had to learn was English intonation. Peter kindly taught me over the telephone. All that remained was to memorise the manuscript (Fig. 7.2).

Ceremony at the IIB

In February I received the schedule of the ceremony. It would run from 4:00 to 9:30 PM and my lecture would last for 60 min. But there was no information about the Prince. Maybe his presence was top secret?

Sachiko and I flew to London by Japan Airlines (JAL) on 7 March. We chose JAL because the travel agent warned us that to fly by British Airways was risky because of British involvement in the first Gulf War.

For the ceremony I wore a dinner jacket and Sachiko wore kimono. All the guests wore dinner suits as well. At about 4:40 PM, someone said that the Prince and the President of the IIB, Professor Wensley Haydon-Baillie, had arrived. We were introduced to them, but I forget what we said.

At 5:00 PM my lecture started and I recovered my thoughts, although Sachiko, who was sitting in the front row, looked a little worried. I spoke about extremophiles for about 50 min without having to consult my manuscript (Fig. 7.3).

The first question was asked by Dr. Ellen Ingham, who was Peter's mother and a retired medical practitioner. She was kind enough to say that my lecture had been very interesting and easy to understand. There then followed a great many questions, to my great relief.

I received the gold medal from Prince Michael and Professor Haydon-Baillie (Fig. 7.4). Professor Haydon-Baillie assured me that my lecture had been a great success and that he would like to quote the last section of it in his own lectures.

Dinner

At 7:00 PM we moved to the dining room. The prince sat at the centre of the table, and I sat on his right with Sachiko to my right as we ate. Suddenly the prince clinked his glass with his knife. This was the signal for the toast. All the diners stood up

THE INTERNATIONAL INSTITUTE OF BIOTECHNOLOGY

Please reply to: The International Institute of
Biotechnology
P.O. Box 228
Canterbury
Kent CT2 7YW

DJH/jmw   11th October 1990

Professor Koki Horikoshi
Dept. of Bioengineering & Bioscience
Faculty of Engineering
Tokyo Institute of Technology
2-12-1, Ookayama
Meguro-ku
Tokyo 152

Dear Professor Horikoshi,

ANNUAL MEDAL LECTURE: 1991

On behalf of The Institute's President and Board of Directors, I am delighted in invite you to deliver the 1991 Annual Medal Lecture in London at The Royal Society on Monday, 11th March 1991.

I am enclosing information on the activities of The Institute from which you will note that a lecture in the area of screening and isolations for biological novelty (including extremophiles and superbugs) is particularly appropriate in view of the recent launch of an Institute search and discovery project under the title of Biodiversity for Biotechnology Innovation. The Medal Lecture is presented to an audience of fairly wide backgrounds and interests, so we ask the Lecturer to cater for experts by providing a forward-looking consideration of the subject whilst at the same time having an eye to the science and public affairs implication of their subject for the more general audience.

If you accept our invitation I will liaise with you on the detailed arrangements for your visit and the Lecture but at this juncture I would say that the Lecture including time for a few questions will occupy approximately one hour. Following the Lecture the President, Professor Wensley Haydon-Baillie, will present you with The Institute's gold Medal and a cheque after which you will be the guest of the President at the Fellows' Dinner in The Royal Society at which you will have the fellowship of The Institute conferred upon you. The Institute makes provision for the Medal Lecturer's travel and accommodation as necessary.

I hope that you will honour The Institute by agreeing to deliver the 1991 Lecture which will the first presented by a Japanese scientist. Please do not hesitate to telephone or fax me if you have any immediate queries. I look forward to receiving your reply.

Yours sincerely,

Dr. David J. Hardman
Secretary

Enc.

A Registered Charity Number 287403. A Company Limited by Guarantee Number 1712316
Registered Office : 165 Queen Victoria Street, London EC4V 4DD, UK
Secretariat address : The International Institute of Biotechnology, P. O. Box 228, Canterbury, Kent CT2 7YW, UK
Telephone (0)227 472099 Fax : (0)227 463482

Fig. 7.1 Official invitation letter from the International Institute of Biotechnology (IIB)

David Harden   Prof. Wenalay Haydon-Baillie   alan T. Bull
Gillian

Michael
Omichead

Your Royal Highness The Prince of Kent, President of The Internation
Institute of Biotechnology, Mr. Chairman, distinguished guests, Ladies and
Gentlemen, it is indeed a great honor for me to be given this opportunity
to present a lecture on biodiversity and biotechnology innovation, with
particular reference to the isolation of superbugs and their industrial
applications.

In autumn 1990, I had the chance to see one of the largest exhibition of
Claude Monet's most famous series, at the Royal Academy of Arts in
London. As Kenneth Clark explains in his famous book "Civilization",
Monet attempted a kind of colour symbolism to express the changing
effects of light. For example he painted a series of cathedral facades in
different lights--pink, blue and yellow-- which seem to me too far from my
own experience. In reality, The colours of these objects depend on the
physical environment, such as sunlight, snow, the time of the day, the
season etc. Under different conditions, one object may show quite
different properties. Who can be sure what is the absolute property? The
microbial world may have the same uncertainty.

It is only three hundred years ago, Antony van Leewenhoek first
observed microorganisms through his microscope. In the middle of the
19th century, Louis Pasteur conducted one of the most important
experiments in the field of microbiology, as a result of which he was able
to refute the theory of spontaneous generation. As you know, Sir
Alexander Fleming made his famous serendipitous discovery of the first
antibiotic, Penicillin in 1928 which has lengthened our average life-span.
The industrial production of Penicillin has resulted in the development of
basic Microbiology, such as physiology and genetics, as well as industrial
microbiology. And in 1977, only one decade ago, the first DNA sequence

1

Fig. 7.2 Manuscript for lecture

Fig. 7.3 Lecturing

Fig. 7.4 After the lecture:
the president of IIB Prof.
Haydon-Bailli (*left*), Koki,
and Prince Michael of Kent
(*right*)

with their glasses, as the prince said: "I ask you all to join me in raising your glass to the health of Professor Horikoshi and his research work." We finished with tea and coffee as usual. Then the prince stood up with Professor Haydon-Baillie, walked to the exit, and got into the professor's car to drive to the prince's residence in Kensington Palace.

I thanked Bill Grant of Leicester University, my friend who had discovered triangular archaea, for coming to the ceremony. He said he had really enjoyed it and was only sorry that he could not bring his wife Susan.

Sachiko had a good opportunity to talk to Professor Alan Bull, of the University of Kent, who was working on antibiotic *Actinomyces*. Sachiko asked him if he had heard of a mythical Japanese tapir that fed on dreams instead of food. "My husband Koki is like that tapir and eats dreams of science. He never talks about money at all." Alan laughed. "My wife Jenny says the same about me, although we did not know about the tapir." Since then they have become family friends.

The next day the court circular in *The Times* reported the award of the gold medal in the presence of Prince Michael of Kent. The guest list was also printed (Fig. 7.5). We still treasure our copy of the newspaper.

Manuscript of Gold Medal Lecture

Introduction

Your Royal Highness Prince Michael of Kent, President of the International Institute of Biotechnology, Professor Wensley Haydon-Baillie, distinguished guests, ladies and gentlemen, it is indeed a great honour for me to be given this opportunity to present a lecture on biodiversity and biotechnology innovation, with particular reference to the isolation of superbugs and their industrial applications.

In autumn 1990, I had the chance to see one of the largest exhibitions of Claude Monet's most famous series at the Royal Academy of Arts in London. As Kenneth Clark explains in his famous book *Civilization*, Monet attempted a kind of colour symbolism to express the changing effects of light. For example he painted a series of cathedral facades in different lights – pink, blue and yellow – which seem to me a long way from my own experience. The colours of these objects depend on the physical environment, such as sunlight, snow, the time of the day, the season, etc. Under different conditions, one object may show quite different properties. Who can be sure what is the absolute property? The microbial world may have the same uncertainty.

It is only 300 years ago that Antony van Leeuwenhoek first observed microorganisms through his microscope. In the middle of the nineteenth century, Louis Pasteur conducted one of the most important experiments in the field of microbiology, as a result of which he was able to refute the theory of spontaneous generation. Alexander Fleming made his famous serendipitous discovery of the first antibiotic, penicillin, in 1928, which has lengthened our average lifespan. The industrial production of penicillin has resulted in the development of basic microbiology, such as physiology and genetics, as well as industrial microbiology. In 1953, Watson and Crick published the double helix structure of DNA. Only two decades later, the first DNA sequence of the virus SV40 was determined by Maxam and Gilbert.

THE TIMES TUESDAY MARCH 12 1991

International Institute of Biotechnology

Prince Michael of Kent was the guest of honour at the Fellows' Dinner of the International Institute of Biotechnology held last night at the Royal Society, and was received by Professor W.G. Haydon-Baillie, president of the institute.

The 1991 Medal lecture was delivered by Professor Koki Horikoshi. The medal and award was donated and presented by Professor W.G. Haydon-Baillie. Professor T. Beppu, Professor H.W. Blanch, Professor C. Bucke, Professor S-T. Chang, Professor R.B. Flavell, Professor A. Goffeau, Professor P.P.Gray, Professor I.J. Higgins, Professor K. Horikoshi, Professor H-J. Knackmus, Dr N.W.F. Kossen, Professor M. van Montagu, Professor M. Reuss, Professor M.S. Salkinoja-Salonen, Professor B. Sikyta, Professor W.D.P. Stewart were admitted to fellowship. Others present were:

Dr S Aaron, Ms K Adams, Dr M Adlard, Dr G W Alderson, Dr and Mrs R Bates, Dr K Beese, Professor A T Bull, Professor T E Burlin, Dr N H Carey, Professor P H Clarke, FRS, Dr R F Coleman, CB, Dr R E Cripps, Dr B Dixon, Dr C S Evans, Professor J George, Dr W D Grant, Professor P Gray, Ms D Haber, Dr and Mrs D J Hardman, Professor G Holt, Professor D A Hopwood, FRS, and Mrs Hopwood, Mrs S Horikoshi, Dr and Mrs D J E Ingram, Professor D P Kelly, Professor Sir Hans Kornberg, FRS, Mrs N W F Kossen, Professor M D Lilly, FEng, and Mrs Lilly, Dr and Mrs C R Lowe, Professor J G Morris, FRS, Mrs M Reuss, Dr B Richards, Sir Denis Rooke, CBE, FRS, FEng, and Lady Rooke, Professor J H Slater, Sir Alex Smith, Mrs W D P Stewart, Dr P J Warren and Dr A Wood.

Fig. 7.5 Next day's *The Times*

Extreme Environments

Not too many years ago, almost all biologists believed that life could survive only within a very narrow range of temperature, pressure, acidity, alkalinity, salinity, and so on, in so-called moderate environments. So when microbiologists looked around for interesting bacteria and other life forms, they attempted to isolate microorganisms only from moderate environments.

Nature, however, contains many extreme environments, such as acidic or hot springs, saline lakes, deserts, alkaline lakes, and the ocean bed. All of these environments would seem to be too harsh for life to survive.

However, in recent times, many organisms have been found in such extreme environments. Moreover, some of them cannot survive in a so-called "moderate" environment. For example, thermophilic bacteria, high temperature-loving bacteria, grow in environments with extremely high temperatures, but will not grow at 20–40 °C. Some alkali-loving bacteria cannot grow in a nutrient broth at pH 7.0, but flourish at pH 10.5. If a moderate environment for conventional organisms such as *Escherichia coli*, which is widely used in the field of genetic engineering, were superimposed on that for thermophilic organisms, for example, the "moderate" environment would seem very cold for thermophiles. Thus, the idea of an extreme environment is relative, not absolute. Clearly we have been too anthropocentric in our thinking. We should therefore extend our consideration to other environments in order to isolate and cultivate new microorganisms.

Discovery of Alkaliphiles

About 23 years ago, I discovered many alkaliphiles.

My interests have been focused on the enzymology, physiology, ecology, taxonomy, molecular biology, and genetics. It is worth noting that alkaliphiles have had a considerable impact in industrial applications. Biological laundry detergents contain alkaline enzymes, such as alkaline cellulases and/or alkaline proteases from alkaliphilic *Bacillus* strains. Another important application is the industrial production of cyclodextrin with alkaline cyclomaltodextrin glucanotransferase. This enzyme reduced production costs and opened new markets for cyclodextrin use in large quantities in foodstuffs, chemicals, and pharmaceuticals.

Superbug Project

The Superbug Project, a 5-year research project of the ERATO program in Japan, was launched in 1984 to search for superbugs that grow in extreme environments, and to try to use their unique properties to establish a "new biotechnology." I was put in charge of the project that was initially called the "Project for studying microorganisms in extreme environments." This is something of a mouthful, so I renamed it the "Superbug Project." It yielded many significant scientific discoveries, some examples of which follow.

Triangular Bacteria

A key event in the superbug saga occurred in 1985, when my colleague, Dr. Bill Grant of Leicester University, found a thin, triangular bacterium living in saltern soil in Japan. No previously known bacteria were triangular – usually they are spheres, rods, or their derivatives.

About 600 samples were collected from various areas of high salt concentration in nature, and small amounts of soil were suspended in a medium containing 20 % NaCl. After 6-day cultivation in the liquid medium at 37 °C with continuous shaking, the cell shape was examined using a phase-contrast microscope. An isolate from a sample of Japanese saltern soil exhibited triangular cells in the liquid medium. The method of cell division of TR-1 attracted our scientific curiosity because of its characteristic shape. The cell division was recorded with a time-lapse cinephotomicroscope equipped with a 16 mm movie camera.

The strain TR-1 is an extremely salt-loving, halophilic archaebacterium. From taxonomic studies, we proposed TR-1 as a new species called *Haloarcula japonica*. We are currently interested in the salt-stable enzymes of this isolate.

Solvent-Tolerant Life Form

From 800 soil samples collected from all over the world, *Pseudomonas putida* IH-2000, a strain that grew well in the toluene medium, was isolated from a soil sample taken in Kumamoto, Kyushu, in Japan, although IH-2000 is unable to utilise toluene as a nutrient. However, benzene, fluorobenzene, lower alcohols, ethers, and ketones exhibit strong toxicity.

Toluene resistance is more than a laboratory curiosity. It has potential uses in industry. For instance, certain fermentation processes such as the bioconversion of steroid hormones require huge amounts of water and extremely large fermentation vessels because of the low solubility of some of the compounds involved. But if we use an organic solvent, we can readily dissolve water-insoluble compounds to effect efficient conversion. This is my philosophy.

Now I should like to talk some more about alkaliphiles.

Studies of alkaliphiles have led to the discovery of many types of enzymes that exhibit many unique properties. In our laboratories about 35 new kinds of enzymes have been isolated and purified. Some of them have been produced on an industrial scale.

Cyclodextrin Production

I would particularly like to highlight cyclodextrin-forming enzymes and the industrial production of cyclodextrin – we call it CD – since the first industrial application of enzymes produced by alkaliphilic bacteria was the production of cyclodextrin from starch.

CD is a derivative of starch and a polymer of seven glucose units. This doughnut-shaped molecule has a hydrophobic cavity. It can be used to make so-called molecular capsules. Volatile compounds are trapped by CD and become nonvolatile, liquids change to powder, and unstable compounds become stable.

The simple method we developed was able to reduce the cost of CD from £400 to £12 per kg. This success has paved the way for its use in a large variety of foodstuffs, chemicals, and pharmaceuticals.

Alkaline Cellulases

No cellulase with an alkaline optimum pH for activity pH 10 or higher had been reported at that time.

About 30 years ago, we found that a newly isolated bacterium produced cellulases which could hydrolyse only modified cellulose in alkaline conditions. This was the first finding of alkaline cellulase in the world.

Several laundry detergents containing the alkaline cellulase are now commercially available. In Japan, such laundry detergents enjoy approximately 60 % of the market. To the best of my knowledge, this is the first-large-scale industrial application of cellulase. I am convinced that no laundry detergent containing alkaline cellulase would have been developed if we had not found and not reported the production of alkaline cellulase by alkaliphilic bacteria.

Conclusion

In closing my talk, I would like to say that a new kind of microbiology has developed very rapidly. This new microbiology is not restricted by the conventional anthropocentric view of microbiology, but is based on studying microorganisms in their optimal conditions for life. We must study them as they are. Seen under the microscope, soil contains 10 million to 1 billion counts of microorganisms per gram. How many of these microorganisms can we isolate and grow? The answer is disappointing. Recovery is only about 1–10 %, even by skilful microbiologists. We have not discovered how to grow all of them, although these microorganisms thrive in nature. This is proof enough that our knowledge is still insufficient. We know even less about extreme environments, but a great variety of these undiscovered microorganisms are distributed on the Earth, indicating the boundlessness of the information they have to offer.

A new 15-year research program, called International Deep Star (Deep-Sea Environment Exploration Program, Suboceanic Terrain Animalcule Retrieval) was launched in October 1990. I have been placed in charge of this program and would like to expand the sources of microorganisms for study from the surface of the globe to the deep sea. We have the use of two submarines, and we can dive to 6,500 m using the Shinkai 6500 of the Japan Marine Science and Technology Centre to collect samples from the deepest parts of the oceans.

Many microorganisms in the deep sea are true superbugs, because they may experience extreme conditions of low temperatures, high temperature, high pressure, or high concentrations of inorganic compounds. Imagine: they have never experienced solar energy, and they have to eat foodstuffs not derived from sunlight. Some of them have an utterly different metabolic pathway from conventional life forms and metabolise inorganic sulfur or iron as their energy source. It is distinctly possible that very ancient life forms may be in hibernation on the seabed, the world's largest refrigerator. Genetic engineering, protein engineering and the so-called modern biology of microbes isolated from the deep sea will give us new information on the origins of life and its evolution.

Finding new life forms will definitely develop basic science and new biotechnologies. Basic science is the one common language of all human beings. We have just started to communicate with nature by using basic science.

Science is just a blank sheet of paper. If Monet placed his colours on this paper, the paper would become a painting. If Beethoven wrote on the paper, the paper would become music. In conclusion, I am convinced that microbiologists will have the opportunity to create a new biotechnology, if they know how to ask the microorganisms.

Thank you for your attention.

Chapter 8
JAMSTEC Deep Star

At the beginning of winter 1989 I met a vice-minister from the Science and Technology Agency who told me that he wanted to set up a deep-sea microbiology research group in the Japan Marine Science and Technology Center (JAMSTEC), and asked for my help (Fig. 8.1). As I was interested in deep-sea extremophiles, I told him that I would be happy to help after 1990, because the ERATO project was planned to end in August. He knew that I had set up the new biology department at the Tokyo Institute of Technology.

I knew that nobody was working on biology in JAMSTEC so I devised a plan to have a clear understanding of what we were going to do.

This deep-sea microbiology research group was to be called Deep Star (Deep Sea Environment Exploration Programme, Suboceanic Terrain Animalcule Retrieval). It was vital to be among the top echelon of deep-sea microbiology.

Because I had no adviser in deep-sea microbiology, I had to feel my own way during a first term of 8 years and a second term of 7 years, totalling 15 years. We adapted the ERATO Superbug Project just finished but slightly modified.

1. We would isolate deep-sea microbes using our three deep-sea submersibles, Shinkai 6,500, Inside of Shinkai 6,500, and Kaiko 10,000 (Figs. 8.2, 8.3 and 8.4).
2. To attract the attention of the mass media to our research, I wanted to show how deep-sea microorganisms could have useful industrial applications.
3. Molecular biology should be incorporated as soon as possible.
4. Physics and physical chemistry should be merged with biology, and we should create an entirely new research centre for deep-sea biology.

   In the spring of 1990, just before starting Deep Star, I visited several researchers in marine microbiology and asked for their future collaboration. However, some researchers were worried about JAMSTEC itself, because the centre had a record of copying others' work. Professor Holger Jannasch of Woods Hole told me: "I do not trust JAMSTEC and I don't think we can work together." I replied:

© Springer Japan 2016
K. Horikoshi, *Extremophiles*, DOI 10.1007/978-4-431-55408-0_8

Fig. 8.1 Location of Japan Marine and Technology Center (JAMSTEC)

Fig. 8.2 The Shinkai 6,500 submersible can go to a depth of 6,500 m

"I have never copied others' scientific data and I do not believe that copying is science." I talked with Holger for almost half a day and in the end we understood each other. I still have his final letter written just a few days before he passed away from cancer (Fig. 8.5).

Fig. 8.3 Inside of Shinkai
6,500 (©JAMSTEC)

Fig. 8.4 The Kaiko 10,000
submersible can go to a depth
of 11,000 m

5. Researchers

   I had six young researchers and three technicians in October 1990, although
   there were no research facilities (Fig. 8.6). JAMSTEC after 2004 is shown in
   Fig. 8.7.

6. Instruments and analysers

   Almost all our cultivation machinery, DNA analysers, and research instruments were brought from RIKEN and ERATO.

7. Facilities

   There were no research facilities until 1992 so I used my research laboratories
   at RIKEN for Deep Star. This collaboration strongly promoted our work in a
   number of fields, especially molecular biology.

From: Holger Jannasch, 67 Church Street, Woods Hole, MA 02543
August 3, 1998.

Dear Friends, *Koki*

This will be necessarily a group-letter. I just want to report the same thing to all of you. I know that some of you were wondering why I dissappeared so suddenly from the lab at the beginning of July. Others have sent get-well cards without really knowing what the matter is all about. Others haven't heard anything yet but might like to know. Well, here it is:

My lymphoma (that, as most of you know, showed up in 1995 and was treated by chemotherapy and radiation) had plateaued all through 1997 and allowed me to travel and do most of my usual activities. The latter included getting ready for and participating in an NSF-funded ATLANTIS/ALVIN cruise during April-May 1998 (We all, 21 microbiologists that is, got all planned samplings, and more. A very successful cruise. Andreas Teske was co-chief scientist).

"Then, during late June, my lymphoma came back again, as the non-Hotchkins-type always does at some unexpected time. And it came back full force. The cat scan showed that most of the lymph system in the abdomen was affected. As a consequence, one of my legs and the abdomen are swollen with fluid which, at times, is very painful. Some fluid went into the lung which makes for very shallow breathing. It is a little better at the moment but, I tell you, being that sick is hard work - and nothing to show for it.

The scary part is that no one knows how long this may go on. I have excellent doctors, but the choices for treatment are very limited. But what I have said in the last sentence of my chapter in the Ann. Rev. of Microbiology will always be true - all sentimentality put aside. From time to time I feel quite comfortable, and then I am reading. I have plenty of books that I wanted to read for years. Friederun is reasonably well herself at the moment and she is taking care of me as no one else could. Her health is more important than mine. Many friends offered their help.

As you see I can even write, at least sometimes and then not more than one letter a day. My son Hans (who was just here with his family for vacation) with the help of B. L. Owens at WHOI got me an old lab top that I can use in the bed. Stephen Molyneaux and Ethel LeFave will be so kind to communicate this note to you - not per bulletin board but as to as many individuals of you as possible. If I forgot one or the other, please forgive me. Then someone will show you this note. If anything, you might experience a little grateful sensation, how good it is to be healthy. Please don't worry about my being sick or feel you have to do something about it. Just put this information in a "sleep" mode somewhere in your head so you don't need any rumors.

With lots of time on hand, I am thinking of each of you very individually.

Yours,
Holger

Last resort: monoclonal antibody treatment. Thanks for your friendship.

Fig. 8.5 The last letter from Prof. Holger Jannasch (Woods Hole)

Fig. 8.6   Tentative building

Fig. 8.7   JAMSTEC (2004)

Results of First-Term Research

Research itself should be original and novel, otherwise I do not believe it can be called research. By 1994 we had isolated new microbes from the deep sea and focused on their applications and molecular biology (Horikoshi and Grant 1998).

1. The isolation and applications of organic solvent-tolerant microbes. It took 2 years to isolate an organic solvent-tolerant bacterium in 1989. However, in 1991 our group developed a sophisticated method to isolate organic solvent-tolerant bacteria from deep-sea samples. Using this method, the isolation frequency of solvent-tolerant microorganisms was about 100 times higher than that of the isolation previously reported. Some of these microorganisms can completely digest crude oil in seawater. This finding has paved the way to removing oil pollution. These organic solvent-tolerant microorganisms have considerable potential for application in bioreactors as solvent-tolerant microbial catalysts.

2. Hyperthermophiles. More than 50 strains of hyperthermophiles have been isolated from hydrothermal vents in the Okinawa area. One of these hyperthermophiles is *Pyrococcus horikoshii*, isolated in collaboration with the Center for Marine Biotechnology at the University of Maryland. The whole chromosomal DNA sequence of this strain was determined by the National Institute of Technology and Evaluation in 1998. Several enzymes in hyperthermophiles have been investigated for industrial applications.

3. Psychrophiles, bacteria that grow at very low temperatures. These bacteria are a good resource for isolating novel enzymes that are active at lower temperatures.

4. Barophiles, bacteria that grow only under hydrostatic pressure.

    Deep Star researchers have succeeded in isolating many obligately barophilic bacteria that function only under high pressure and have isolated the genes responsible for barophily. Phylogenetic studies of the isolates suggested that the isolated strains belong to a new sub-branch of the gamma proteobacteria group. It was discovered that the *lac* promoter was activated under high pressure in the absence of inducers. Although the enzymes produced by these barophiles have not been studied, it is highly likely that it will be possible to clone novel genes from barophiles and to express them in conventional microorganisms

5. Work is proceeding on preservation and culture collection of deep-sea microorganisms.

Review of Deep Star at JAMSTEC

I attended the review panel's interview on 9–10 November 1995. The head reviewer was Professor Rita Colwell (Fig. 8.8) of the University of Maryland. (She served as the 11th Director of the National Science Foundation from 1998 to 2004.)

Fig. 8.8 Professor Rita
Colwell (University of
Maryland)

After the interview, she kindly gave me a draft of the panel's review, which I was happy to receive. A part of the draft read as follows:

"Dr. Koki Horikoshi was the director of the ERATO Superbugs Project and very much a traditional leader in that he was older and more established. His project was to look at unusual microorganisms, particularly bacteria that grow at very high pH (12–14, close to absolute basic pH), such as alkaliphilic bacteria that grow in solvents. Horikoshi actually isolated a bacterium that grew in practically full-strength toluene, for example. It was an interesting feat that took 2 or 3 years.

The Superbug Project was a reasonably successful ERATO project, in that some new, young scientists had been trained and several successful patents had come out of the project. For example, the bacteria that were capable of operating at very high pH and had a proteolytic activity were utilised in a very popular detergent in Japan. So there were successful applications of this ERATO programme.

What is interesting is that, subsequently, Horikoshi was selected to lead a Superbug continuation project in Yokosuka through JAMSTEC. Five scientists were recently asked to carry out a review of that programme: Dr. Harlyn Halverson from the University of Massachusetts, Dr. Ian Dundas from the Bergen Rienvow Tech Center in Norway, Dr. Rita Colwell from the University of Maryland Biotechnology Institute, and two researchers from Japan: Dr. Koichi Oada, who heads the Ocean Sciences Research Institute of the University of Tokyo, Marine Microbiology Group, and Dr. Tadashi Matsunaga of the Tokyo University of Agriculture and Technology. This team reviewed Horikoshi's JAMSTEC project on 9–10 November 1995.

The programme at Yokosuka turns out to be rather extraordinary in that $5–20 million a year has been expended since 1991. At this JAMSTEC project there are 29 researchers working on deep-sea research; and from 1992 to 1995 there were also 21 researchers from six countries who came to work with them. Engineers and scientists at this project spent 2 years constructing an incredible facility that allows them to collect microorganisms under deep-sea hydrostatic pressure and grow them continuously in culture, transferring them constantly under pressure. These scientists also have developed the capability of working at very high temperatures, to better preserve and study deep-sea hydrothermal vent bacteria.

Researchers at the Yokosuka JAMSTEC programme called Deep Star have built two submersibles. In addition to the Shinkai 2000, which allowed them to go to 2,000 m, they have built the Shinkai 6500, which will allow manned (human-operated) descent to the deepest trenches of the world's oceans, to a depth of 6,500 m.

The Deep Star project is unique in that it targets only deep-sea areas and microorganisms in that environment. It is an extension of Horikoshi's Superbug Project, but it goes into its focus very extensively, to the very deep part of the ocean: this has necessitated the development of a considerable amount of expensive and unique hardware.

Five groups in the JAMSTEC Deep Star project are focused on organic, solvent-tolerant microorganisms. This is the extension of Horikoshi's original ERATO project, with the addition of hyperthermophiles, bacteria that grow at temperatures of 110–140° or 150 °C under high pressure; psychrophiles, bacteria that grow at very low temperatures (this, of course, is a characteristic of the deep sea); and barophiles, bacteria that grow only with hydrostatic pressure. JAMSTEC researchers have succeeded in isolating two obligatory barophilic bacteria that function only under high pressure.

Basically, the Deep Star programme has developed a unique set of engineering skills for the construction of submersibles and of isolation and incubation chambers for research. These engineering capabilities are not matched anywhere else in the world. It is clear that the next phase of the creative basic research work that spun off from ERATO provides a capability and commitment for much longer support, and this has enabled the group to devise an apparatus for very precise sampling of microorganisms and the capability of studying the properties of microorganisms under natural conditions.

The Deep Star project was possible only because of Horikoshi's previous experience and equipment; but as the new facilities are now available, expansion into molecular techniques has been introduced. This expansion represents the next phase and an opportunity for young scientists. One of the very strong recommendations made by the review team is that in the next round, the next 5 years, leadership be initiated by Horikoshi, but that one of the "young tigers" aged 35–40 be groomed to take over. Unfortunately, there is not any one on site among the younger workers who can take on this leadership. That seems to be a shortcoming, and it may well be a shortcoming of the ERATO projects in general, at least of the older ones, not to have a capacity to cultivate the next head or director from within the group.

The JAMSTEC Yokosuka facility is the premier centre for this kind of research in the world. The only other related programmes are Ifremer in France and Woods Hole in the United States. But these are secondary; they really do not have the capacity, the skills, and the breadth of the JAMSTEC Deep Star programme.

The 5-year-old Deep Star project has obtained good results. For example, new selective procedures have resulted in the isolation of a number of solvent-tolerant strains in mutants. The mutants are the next step, because solvent-tolerant strains came from the ERATO technology. What are particularly interesting are the barophiles that have been isolated, adding to the very meagre knowledge of these organisms, and the molecular genetic studies of baro-tolerance, especially in the identification of a very specific protein that appears to be correlated with the capacity to grow under pressure. The Deep Star staff have begun studies of the psychrophiles, the bacteria capable of growing at very low temperatures, around 0–10 °C, and they are looking at the interrelationships of temperature, salinity, pressure, and other environmental parameters. The results are at a very early stage but constitute a sound basis for launching this major research programme on systematics, biodiversity, and preservation methods.

Horikoshi has provided strong leadership. He has brought in scientists from within Japan and from other countries. The key factors have been the flexibility and recruitment aspects of the programme. The publication record has been very good. The Deep Star programme researchers had published 24 original papers by 1994. By the end of 1995 they had published about 56 papers in total, and these papers have been published in very good journals. Researchers at the Deep Star programme have also made 132 presentations around the world.

At this stage of development of the programme the review committee concluded that a very healthy beginning has been made for Deep Star, but that in the next phase several factors will be critical. Every effort, this panel believes, must be made to increase collaboration with universities and other research centres in Japan and abroad; to increase communication with the disciplines that are involved – engineering, molecular biology, protein chemistry and so forth; and to maximise opportunities to interact with related basic sciences."

Chapter 9
Samples from 10,898 m Beneath the Pacific Ocean

On 28 February 1996 Japanese researchers working on the Deep Star programme broke their own world record by sending the unmanned submersible *Kaiko* (the Japanese word for "trench") to a depth of 10,898 m beneath the Pacific Ocean (Fig. 9.1). *Kaiko* sent back video images of life in the depths of the Mariana Trench, the deepest point in the world's oceans. A brief but excited fax from the mother ship described the scene captured on *Kaiko*'s camera: "The bed of the Mariana Trench was filled with a fine mud of reddish brown (Fig. 9.2). There were no rocks or cracks at all and it resembled a desert. However, very unusual organisms were observed here and there." In the following fax, they described types of sea urchin, quite fast-moving jellyfish, and the excrement of sea organisms. My colleagues also saw a fast-moving shrimp about 3 cm long (Figs. 9.3 and 9.4). There was also a kind of sea cucumber, which was the same size as the jellyfish. *Kaiko* scooped up samples of mud to bring to the surface (Kato et al. 1997; Kato et al. 1998). Now we would be able to isolate living creatures from the deepest point of the Mariana Trench (Table 9.1).

In an attempt to characterize the microbial flora on the deepest sea floor, we isolated thousands of microbes from a mud sample collected from the Mariana Trench (Takami et al. 1997). The microbial flora found at a depth of 10,898 m was composed of actinomycetes, fungi, non-extremophilic bacteria, and various extremophilic bacteria such as alkaliphiles, thermophiles, and psychrophiles. Phylogenetic analysis of Mariana isolates based on 16S rDNA sequences revealed that a wide range of taxa were represented (Horikoshi and Grant 1998).

I told Peter Ingham about these results and he kindly introduced me to Anjana Ahuja, science correspondent of *The Times*. On 18 March *The Times* carried her report under the headline "Faces from the final frontier." Shortly after its publication, this article was to have an unexpected consequence.

In 2012, less expensive and simpler unmanned deep-sea research devices jointly developed by backstreet factories in Tokyo took pictures of fish nearly 8,000 m below the sea surface in the Japan Trench.

© Springer Japan 2016
K. Horikoshi, *Extremophiles*, DOI 10.1007/978-4-431-55408-0_9

Fig. 9.1 *Kaiko* 11,000

Fig. 9.2 The seabed of the Mariana Trench

Fig. 9.3 Shrimps living at depth 10,000 m

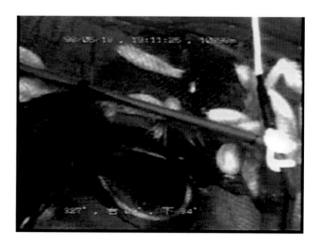

Fig. 9.4 Shrimp
(*Hirondellea gigas*), about
3 cm in length

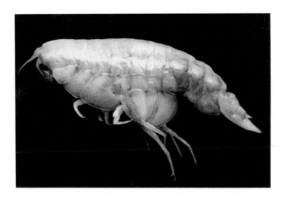

Table 9.1 Extremophiles isolated from the deepest sea mud of the Mariana Trench

Category	Isolation conditions	Bacteria recovered (colonies g⁻¹ dry sea mud)
Alkaliphile	pH 9.5–10.0 25 °C, 0.1 MPa	4.1×10^2–1.2×10^3
Barophile	100 MPa 25 °C, pH 7.6	–
Thermophile	55–75 °C pH 7.6, 0.1 MPa	5.8×10^2–3.5×10^3
Psychrophile	4 °C pH 7.6, 0.1 MPa	2.0×10^2
Halophile	15 % NaCl, 25 °C pH 7.6, 0.1, 100 MPa	–
Acidophile	pH 3 25 °C, 0.1, 100 MPa	–
Non-extremophile	25 °C, 0.1 MPa pH 7.2±0.4	2.2×10^4–2.3×10^5

– no growth obtained

Three "Edokko No. 1" submersibles (Fig. 9.5) were released by the survey ship *Kaiyo* about 200 km off the coast of the Boso Peninsula in Chiba Prefecture between November 21 and 23.

The devices, each weighing 50 kg and equipped with three-dimensional (3-D) video cameras, were developed by small family-owned factories in the "shitamachi" area of old Tokyo. They were dropped to depths as great as 7,860 m, and two succeeded in photographing a school of eel-like deep-sea fish (Fig. 9.6).

Even now, the researchers looking at microorganisms in the mud samples have to culture them at a pressure of 1,000 atm. Some microorganisms can even be grown at more than 1,000 atm. Recently, we found that living organisms can be surpris-

Fig. 9.5 Three "Edokko No.1" submersibles, which are simpler and less expensive, seen as deep as 8,000 m

Fig. 9.6 Eel-like deep-sea fish at 7,860 m

ingly proliferative during hyperacceleration (Figs. 9.7 and 9.8). In tests, a variety of microorganisms, including gram-negative *Escherichia coli*, *Pyrococcus denitrificans*, and *Shewanella amazonensis*; gram-positive *Lactobacillus delbrueckii*; and eukaryotic *Saccharomyces cerevisiae*, were cultured while being subjected to hyperaccelerative conditions (Abe and Horikoshi 1998). We observed and quantified robust cellular growth in these cultures across a wide range of hyperacceleration

Microbial growth at hyperaccelerations up to 403,627 × *g*

Shigeru Deguchi (出口 茂)[a,1], Hirokazu Shimoshige (下重裕一)[a,2], Mikiko Tsudome (津留美紀子)[a], Sada-atsu Mukai (向井貞篤)[a,b], Robert W. Corkery[c], Susumu Ito (伊藤 進)[d], and Koki Horikoshi (掘越弘毅)[a]

[a]Institute of Biogeosciences, Japan Agency for Marine-Earth Science and Technology, Yokosuka 237-0061, Japan; [b]Institute for Advanced Study, Kyushu University, Higashi-ku, Fukuoka 812-8581, Japan; [c]Institute for Surface Chemistry, 114 28 Stockholm, Sweden; and [d]Department of Bioscience and Biotechnology, Faculty of Agriculture, University of the Ryukyus, Nishihra, Okinawa 903-0213, Japan

Edited by Henry J. Melosh, University of Arizona, Tucson, AZ, and approved March 15, 2011 (received for review December 19, 2010)

It is well known that prokaryotic life can withstand extremes of temperature, pH, and radiation. Little is known about the proliferation of prokaryotic life under conditions of hyperacceleration attributable to extreme gravity, however. We found that living organisms can be surprisingly proliferative during hyperacceleration. In tests reported here, a variety of microorganisms, including Gram-negative *Escherichia coli*, *Paracoccus denitrificans*, and *Shewanella amazonensis*; Gram-positive *Lactobacillus delbrueckii*; and eukaryotic *Saccharomyces cerevisiae*, were cultured while being subjected to hyperaccelerative conditions. We observed and quantified robust cellular growth in these cultures across a wide range of hyperacceleration values. Most notably, the organisms *P. denitrificans* and *E. coli* were able to proliferate even at 403,627 × *g*. Analysis shows that the small size of prokaryotic cells is essential for their proliferation under conditions of hyperacceleration. Our results indicate that microorganisms cannot only survive during hyperacceleration but can display such robust proliferative behavior that the habitability of extraterrestrial environments must not be limited by gravity.

astrobiology | extremophiles

The robustness of prokaryotic life to physical extremes of temperature, pH, pressure, and radiation is well known (1) and has led to their ubiquitous presence on Earth (2, 3). Resilience to physical extremes is also extremely likely to be required for the existence of life beyond this planet (1, 4). Finding extraterrestrial life is a major motivation driving searches for extrasolar planets (5); thus, understanding the physical limits for known organisms is crucial in evaluating the probability that such planets harbor life (1, 4). Assessing the habitability of extraterrestrial environments requires an expanded set of criteria involving factors that can be ignored for terrestrial environments. The effect that extremes of gravity have on organisms is one such factor to consider when exploring for life beyond Earth.

The effect of microgravity on biological processes has been an active area of research particularly because it is relevant to human health during space flight (6, 7). Microorganisms make ideal model life forms for microgravity research because they are lightweight, small, and relatively easy to handle in space and have short generation times (7). Consequently, numerous experiments have been performed on microorganisms both in orbit and on Earth-based clinostats that simulate microgravity. The results demonstrate that microgravity affects microorganisms in a wide variety of ways related to their growth, physiology, pathogenesis, stress resistance, and gene expression (7–23).

The majority of these studies indicate that microgravity stimulates the growth of microorganisms (e.g., *Salmonella enterica* serovar Typhimurium, *Bacillus subtilis*, *Escherichia coli*) compared with 1 × *g* controls (8–17). In the case of *E. coli*, for example, the lag phase was shortened, the duration of exponential growth was increased, and the final cell population density was approximately doubled during space flights (13). Simulated microgravity can also affect the secondary metabolism of microorganisms. For example, production of β-lactam antibiotics by

Streptomyces clavuligerus, production of rapamycin by *Streptomyces hygroscopicus*, and production of microcin B17 by *E. coli* were suppressed during culturing in simulated microgravity, whereas production of gramicidin S by *Bacillus brevis* was unaffected (14, 18). *S. enterica* serovar Typhimurium showed enhanced virulence in a murine infection model (19, 20) conducted in space flight and under modeled microgravity compared with conditions of normal gravity (19, 20). These microorganisms also showed increased resistance to environmental stresses, increased survival in macrophages, and significant changes in protein expression levels (19). To elucidate the molecular mechanisms of microbial responses to microgravity, 2D gel electrophoresis and DNA microarray analysis have been used (19–23). Recent analysis of *S. enterica* serovar Typhimurium grown in space identified 167 transcripts and 73 proteins that changed expression compared with ground controls, and conserved RNA-binding protein Hfq was identified as a likely global regulator (20). Gene expression of eukaryotic *Saccharomyces cerevisiae* is also affected by simulated microgravity (22, 23).

Compared with the relatively active research on microbial responses to microgravity, there are fewer studies that report experiments on microorganisms exposed to gravities greater than 1 × *g* (11, 16, 24–28). Unlike in microgravity, experiments in hypergravity were performed exclusively in simulated environments and primarily by subjecting microorganisms to centrifugal acceleration in centrifuges. Bouloc and D'Ari (11) reported that hyperaccelerations of 3 and 5 × *g* did not affect the growth of *E. coli*, whereas Brown et al. (16) observed growth suppression at 50 × *g*. Similar observations were reported for *Paramecium tetraurelia*, which showed no effect at 10 × *g* but a significantly lower proliferation rate and lower population density at 20 × *g* (24).

At hyperaccelerations much greater than ~10² × *g*, the effect of sedimentation on microbial cells becomes significant. In a typical example, cultures of bacterial cells subjected to centrifugation at 3,000–5,000 × *g* for 5–10 min yielded pellets of intact bacterial cells (29). If microbial growth had occurred under these (or similar) conditions, it must have happened within or on the pellet. In stark contrast, the effect of cellular sedimentation is not very significant at lower accelerations, where growth can occur planktonically. Studying microbial proliferation, not simply survival, at such hyperaccelerations addresses the fundamental biological question of what are the physical limits of organismic viability (1) under a range of gravitational accelerations larger than those found on Earth. Understanding the gravity limits for microorganism growth

Author contributions: S.D., H.S., M.T., S.I., and K.H. designed research; S.D., H.S., M.T., and S.M. performed research; S.D., S.M., and R.W.C. analyzed data; and S.D. and R.W.C. wrote the paper.

The authors declare no conflict of interest.

This article is a PNAS Direct Submission.

[1]To whom correspondence should be addressed. E-mail: shigeru.deguchi@jamstec.go.jp.

[2]On leave from: Graduate School of Interdisciplinary New Science, Toyo University, 2100 Kujirai, Kawagoe 350-0815, Japan.

This article contains supporting information online at www.pnas.org/lookup/suppl/doi:10.1073/pnas.1018027108/-/DCSupplemental.

Fig. 9.7 Microbes can grow at hyperacceleration of 400,000 g

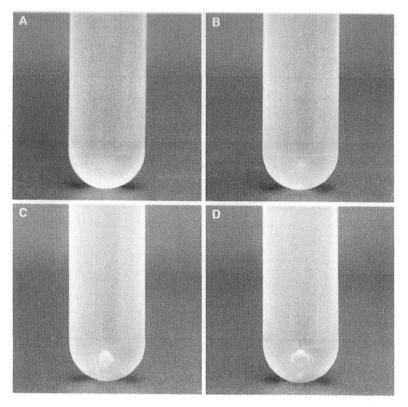

Fig. 9.8 Growth under 400,000 g and 30 °C after (**a**) 0 h, (**b**) 6 h, (**c**) 24 h, and (**d**) 48 h

values. Most notably, the organisms *P. denitrificans* and *E. coli* were able to proliferate even at 403,627 g (Deguchi et al. 2011). Our results indicate that microorganisms not only can survive during hyperacceleration but can display such robust
proliferative behaviour that clearly the habitability of extraterrestrial environments
need not be limited by gravity.

Chapter 10
His Majesty the Emperor and the Deep Sea

Lecture to the Emperor on Organisms in the Mariana Trench

On April 2, 1996, I received a phone call from Chamberlain Katsusuke Meguro of the Imperial Household Agency. He told me: "His Majesty The Emperor read *The Times* of 18 March and would like to know more about the organisms found at 10,000 m in the Mariana Trench." The Emperor is famous as a biologist. I said that I would be honoured to come and asked when the best time might be as I had to attend the International Meeting for Extremophiles in Portugal at the end of May. Finally we arranged the meeting for July 2 at 3 pm at the Fukiage Palace (Private Palace). Sachiko asked why I had imposed my schedule on the Emperor. Chamberlain Meguro said that the Emperor always made arrangements to suit the scientist's convenience.

After the meeting in Portugal, I prepared a lecture with videotapes, photographs, and samples. At 2 PM I went to the palace and was shown into a waiting room. After a short while, Chamberlain Meguro came and announced that the Emperor was waiting for me. In the meeting room the Emperor was standing at the door, and Chamberlain Meguro introduced me.

Lecture to the Emperor

Research into deep-sea microbes started about 60 years ago, and their taxonomies were major works. Molecular biology and molecular genetics had not been introduced by 1990. When I accepted the offer to establish a biology section at JAMSTEC, I decided to change the classic deep-sea microbiology to modern biology using genetic engineering techniques.

We would like to understand the origins of life, if we can. The deep seas offer extreme conditions, such as higher pressure, high temperatures, and low temperatures too. It is quite possible that microorganisms have been dormant and unchanging

© Springer Japan 2016
K. Horikoshi, *Extremophiles*, DOI 10.1007/978-4-431-55408-0_10

for 3.5 billion years in the world's largest refrigerator. We hoped to gather new information on old microbes. Furthermore, we hoped to be able to open new industrial applications using the genes of these microbes.

Using two submersible vessels (Shinkai 2,000 and Shinkai 6,500), we were able to see cod and sea cucumbers in the deep sea and to collect samples from the deep ocean floor. Numerous microorganisms – thermophiles, psychrophiles, alkaliphiles, and conventional microbes – were isolated from these samples. The most characteristic microbes are piezophiles, some of which could not grow at 1 atmosphere but exhibited good growth at 500 atm. We found pressure-dependent promoters in them. These results indicated that we have to cultivate the microorganisms in the cultures in which they live.

On 28 February 1996 Japanese researchers working on JAMSTEC Deep Star program sent the unmanned submersible *Kaiko* to 10,898 m beneath the Pacific Ocean. *Kaiko* sent back video images of life in the depths of the Mariana Trench, the deepest point in the world's oceans.

We must learn about the cultures of microbes, and then cultivate them in their own environments.

After the lecture I returned to home with a souvenir, pancakes with bean jam inside. It was a great honour for me to be invited to such an illustrious meeting.

King Charles II Medal

On 18 May 1998 Sachiko and I received an invitation from the Royal Society in London to attend the presentation on 28 May of the Royal Society's King Charles II medal to Emperor Akihito of Japan (Fig. 10.1).

At a special ceremony, in front of a packed hall at the Royal Society, the Emperor received the Medal from the president of the Royal Society, Sir Aaron Klug (Fig. 10.2).

The King Charles II Medal was created to give recognition to foreign heads of state who have made an exceptional contribution to the promotion of science and its place in society. On the face of the medal (about 7 cm in diameter) is the head of King Charles II, and three lions and the Latin words *Gloria principum rem investigate* are on the reverse.

The Emperor is an active scientist, who has contributed greatly to the understanding of fish biology and taxonomy and has published extensively, particularly on the *gobi* fish of Japan, and he also takes a keen interest in promoting the conservation of endangered species. The Emperor is also active in the support and promotion of science in Japan through his involvement in the meetings and ceremonies of the Japan Academy and the major Japanese international scientific awards.

18-05-98 09:52   FROM: TO:90081339940438   PAGE:01

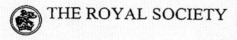

THE ROYAL SOCIETY

6 CARLTON HOUSE TERRACE LONDON SW1Y 5AG
Tel: 0171-839 5561 Fax: 0171-451 2691
DDI: 0171-451 2503   email: Tony.Leaney@royalsoc.ac.uk

FACSIMILE TRANSMISSION

To: *Professor Horikoshi*   Date: *18/5/98*

At: Fax No: *81 339940438*

From: *Tony Leaney*

Please reply to:

Pages (excluding this sheet):

MESSAGE

*On the Occasion of the Award of the King Charles II Medal
To His Majesty Emperor Akihito of Japan*

*The President of the Royal Society
Sir Aaron Klug O. M. F.R.S.
has pleasure in inviting*

Professor Kouki & Mrs Horikoshi

*to the Ceremony and Reception at
6 Carlton House Terrace on
Thursday 28 May 1998 at 2.00p.m. for 2.30p.m.*

To remind.
The Royal Society (ref. CAS)
6 Carlton House Terrace
London SW1Y 5AG
Tel: 0171-451 2503

Guests are asked to be
seated by 2.30 p.m.

Please bring this
admission card with you

Registered Charity No. 207043

WWW address: http://www.royalsoc.ac.uk

Fig. 10.1 The official invitation letter from The Royal Society

Fig. 10.2 The Emperor
received the King Charles II
Medal from Sir Aaron Klug

The Emperor's Acceptance Speech

The Emperor's acceptance speech to the Royal Society, which he delivered in impeccable English, was as follows (Fig. 10.3):

It is now more than 300 years since King Charles II granted the Royal Society its first charter. I would like to express my profound respect for the society, for the unceasing contributions it has continued to make towards the development of world science.

In the 1960s, when I started my research on gobioid fish, there were not yet many specialists in the field, and few up-to-date references. I recall the joy I felt when, after days and days of peering into the microscope, I finally came to identify the characters of the arrangements of rows of sensory papillae of one species after another.

By that time, the method of investigating bones by dyeing them with alizarin had already been widely adopted, but little had been done as far as the bones of gobioid fish were concerned. The first bone I myself examined was the scapula. The inspiration for this came partly from a paper by Mr. Tate Regam of the British Museum of Natural History in 1911. In his paper he pointed out that one of the marks distinguishing the family Eleotridae from the family Gobiidae, both of which belong to gobioids, was the existence of the scapula. However, Dr. W.A. Gosline of the University of Hawaii presented an alternative view in his paper of 1955. I myself, after close examination of scapulae of various species, reached the general conclusion that the shape of the scapula is largely the same within one genus and can therefore by taken as a defining character of the genus, though not of a specific family. From then on, I concentrated on osteology and eventually discovered a species of gobioid fish whose bones were not specified. There are more than 300 species of gobioid fish in Japan, and they have diverted

Emperor honoured for contribution to science

At a special ceremony, in front of a packed hall at the Royal Society, the Emperor Akihito of Japan received the King Charles II Medal from the President Sir Aaron Klug. In the audience was His Royal Highness The Duke of Edinburgh and many senior members of the scientific, parliamentary, business and diplomatic communities.

The King Charles II Medal was instigated to give recognition to Foreign Heads of State who have made an exceptional contribution to the promotion of science and its place in society.

The Emperor is an active scientist who has contributed greatly to the understanding of fish biology and taxonomy. He has published extensively, particularly on the gobi fish of Japan and also takes a keen interest in promoting the conservation of endangered species. The Emperor also is active in the support and promotion of science in Japan through his involvement in the meetings and ceremonies of the Japan Academy and the major Japanese international scientific awards such as The Japan Prize, Kyoto Prize and the International Prize for Biology.

> 'The Emperor is an active scientist who has contributed greatly to the understanding of fish biology'

The Emperor's acceptance speech, which he delivered in impeccable English, is reprinted in full opposite.

The award of the medal took place on 28 May 1998 during the Emperor's state visit to the UK.

Following the ceremony the Emperor and Empress were able to meet Nobel Prizewinners, Fellows of the Society and other guests at a reception in their honour. Their Imperial Majesties clearly appreciated the opportunity to mingle with the guests and introduced an air of informality into the occasion which was greatly enjoyed by all those present.

Emperor Akihito of Japan receiving the King Charles II Medal from the President, Sir Aaron Klug

UK and Japan: partners in science

Ahead of the visit by His Imperial Majesty, the Emperor of Japan, the Society held a conference to celebrate the UK-Japan scientific partnership.

The conference, intended to reveal the opportunities available for scientific research in Japan, was attended by 180 delegates, including diplomats and government officials, industrialists, academics, Fellows of the Royal Society, established scientists and young researchers.

The conference began with an overview by the Master of Churchill College, Cambridge, Sir John Boyd, Ambassador to Japan until 1996. Chaired by the Foreign Secretary, there followed discussions and presentations by Professor Hiroyuki Yoshikawa, President of the University of the Air, the JSPS and the Science Council of Japan and Professor Michael Pepper, FRS of the Cavendish Laboratory, Cambridge and the Toshiba Cambridge Research Centre. A more personal perspective, on their experiences in Japan, was given by two Royal Society awardees: Dr Susan Evans, Reader in Comparative Anatomy at University College London and Dr Martyn Kingsbury of the Academic Cardiology Unit, St. Mary's Hospital Medical School.

With the aim of highlighting the benefits to be gained from collaboration, the conference provided a valuable networking opportunity and forum for disseminating information on Japanese institutes. Displays were on view throughout the morning to promote exchanges with Japan and to show through videos, literature and posters a taste of the Japanese language and culture.

Royal Society News – July 199

Fig. 10.3 The Emperor's speech

into various forms. But their bones always show the tendency of degeneration or disappearance. This led me to conclude that the species of gobioid fish in which the bones show the least osteodegeneration must be closest to the common ancestor. I am very much looking forward to seeing the phylogeny and relations of gobioid fish further clarified in the future by the most advanced methods of biological research.

In the course of carrying out my researches, I have had opportunities to communicate with Dr. Peter J. Miller, professor at Bristol University, and many other ichthyological specialists both at home and abroad. I have also had opportunities to borrow, or be given, precious specimens from individuals and various museums, including the British Museum of Natural History. I am most grateful for all this help.

When I was Crown Prince, I was elected a foreign member of the Linnaean Society of London. It was a great honour indeed to be admitted as 1 among only 50 foreign members. Whilst I felt that this was too great an honour for me, it stimulated me to continue my studies so as to maintain the high standards of research worthy of this honour, which remains one of my dearest memories.

Since my accession to the throne, my full daily schedule has so far prevented me from continuing my researches effectively. Indeed, the only paper I have completed since enthronement was one that had been largely done some 10 years before, when I was still Crown Prince. I am firmly resolved, however, not to let my light of learning be extinguished. I hope to go on pursuing my studies, mindful of the fact that I have today been awarded the King Charles II Medal.

Reflecting on the history of science in Japan, one cannot fail to recognise the role played in its development by cross-border cooperation from various people abroad, including many British scientists.

Scientific truth must always be universal.

I would like to conclude this word of thanks by expressing my hope that British–Japanese scientific exchanges will continue to advance, thereby making significant contributions to the progress of world science.

After the Emperor's speech, a reception was held. Their Imperial Majesties the Emperor and Empress talked with a number of people, including us, and I had the opportunity to show them a picture of red shrimp caught in the Mariana Trench (Fig. 10.4). A few months later in Tokyo I took a sample of the red shrimp to the marine biologist Emperor.

Their Imperial Majesties the Emperor and Empress at JAMSTEC and the Japan Academy

On 28 March 2001, their Majesties the Emperor and Empress toured our laboratory at JAMSTEC in Yokosuka. We discussed the phylogenic tree of gobioid fish and red shrimps. Furthermore, the Emperor boarded the Shinkai 6,500, staying almost 30 min in the small sphere (less than 2 m in diameter) and asked several scientific questions (Fig. 10.5).

Fig. 10.4 I had the opportunity to talk about red shrimp of the deep sea

Fig. 10.5 The Emperor and Empress toured our laboratory at JAMSTEC

Fig. 10.6 I received the Japan Academy Prize for my work on alkaliphiles with the Emperor and Empress in attendance

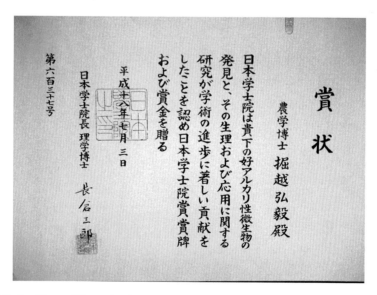

Fig. 10.7 Certificate of Japan Academy Prize (2006)

On 3 July 2006 I received the Japan Academy Prize for my research work on alkaliphiles (Figs. 10.6, 10.7 and 10.8). We had dinner at the Imperial Palace. The Emperor sat next to me and asked about progress in the field of deep-sea exploration.

Fig. 10.8 *From the left*: Asako (wife of Toshiaki), Sachiko, Koki, and Toshiaki at the ceremony

I told him that I was now able to isolate microbes from our deep-sea samples. Almost all the microbes were entirely new. I asked the Emperor if the phylogenetic tree of the DNA genes of gobioid fish had been determined. He replied that it had been completed.

The dinner party was like a mixer at a scientific meeting. It remains a joyous memory.

Chapter 11
New Scientific Journal *Extremophiles* and the International Society of Extremophiles

Launch of a New Scientific Journal, *Extremophiles*

At the end of the 1980s, extremophiles were not a popular topic, and research papers on microbes living under unusual conditions met with a cold reception. Many microbiologists used to joke, "Only abnormal researchers investigate abnormal microbes." In 1995 Ms. Aiko Hiraguchi of Springer visited my office in Shinbashi, Tokyo, and we enjoyed a chat. We were about to stage the First International Symposium on Extremophiles in Estoril, Spain, in 1996. I thought this would be a good time to launch a new journal for biologists working on extremophiles. As the result of this conversation, we agreed to publish an entirely new type of journal through Springer.

I sent a letter to microbiologists all over the world, proposing the launch of a new extremophiles journal. However, it seemed that it was unheard of to create a new scientific journal.

Ignorance is bliss. I phoned about 20 well-known biologists and received positive answers from William D. Grant (UK), Garo Antranikian (Germany), Karl Stetter (Germany), Terry A. Krulwich (USA), and Juergen Wiegel (USA). They agreed to join the journal as managing editors. I then proposed that I should take care of the journal as chief editor.

I attended the international symposium in Estoril, and requested that the Second International Symposium on Extremophiles be held in 1998 at Yokohama, Japan. I then announced to the delegates that we intended to launch a new journal, *Extremophiles*, in 1997. Not only did we not have a scientific society but we had no journal. Therefore, we had to establish an entirely new journal for research into extremophiles, because the progress of science was so rapid.

© Springer Japan 2016
K. Horikoshi, *Extremophiles*, DOI 10.1007/978-4-431-55408-0_11

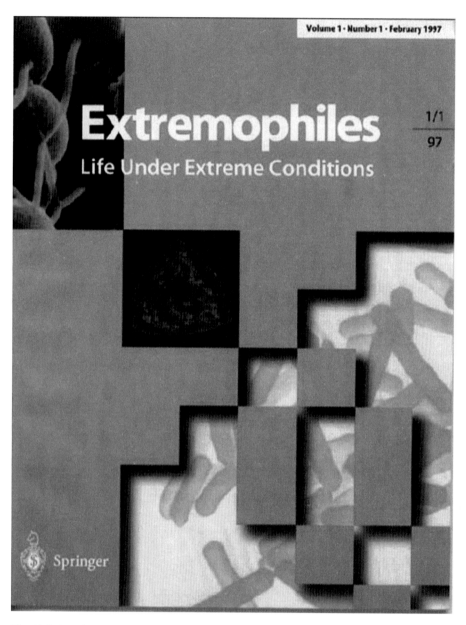

Fig. 11.1   The first issue of *Extremophiles*

In February 1997 the first issue of *Extremophiles: Life Under Extreme Conditions* was published (Fig. 11.1). On the first page, I offered the following editorial:

Extremophiles are organisms that grow in extreme environments such as high temperature, pressure, or salt concentration or at unusual levels of nutrients, pH, oxygen, organic solvents, or heavy metals. In 1974, MacElroy first used the term

in his paper "Some Comments on the Evolution of Extremophiles" to describe organisms able to populate and grow in extreme environments (MacElroy 1974). At that time, extremophiles included the thermophiles, the psychrophiles, the halophiles, and the alkaliphiles. Since then, many interesting extremophiles have been isolated from such environments, and extensive studies have been carried out on some organisms to elucidate the structure of their DNA, RNA, proteins, lipids, and polysaccharides as well as their tolerance mechanisms, metabolic pathways, and gene expression. In some cases enzymes that are stable at high temperature and high pH values have been isolated and commercialised. Extremophiles offer important insights into the biology and evolution of many organisms, and they hold the promise of providing valuable molecules of industrial and biotechnological significance. *Extremophiles* is the first journal to be devoted to these exciting organisms and provides an international forum for basic and applied research on organisms living under extreme conditions.
Koki Horikoshi, Chief Editor, February 1997.

The journal contained one review and five original research papers. I was extremely happy to see that first issue. I still keep three copies in my bookcase as a treasure. The impact factor was 3.13. Since 2008, I have been the Founding Editor and Prof. Garo Antranikian of Hamburg University of Technology has become the Chief Editor.

Second International Congress on Extremophiles 1998 in Yokohama

We held the International Congress on Extremophiles 1998 from January 18 to 22 in Yokohama (Fig. 11.2), and 150 foreign and 200 domestic researchers participated.

The congress was composed of two special lectures, five sessions, and a panel discussion, including not only microbiology on the Earth, but also new fields of exobiology, subsurface microbiology, and genome analysis. We focused on both fundamental and application studies of "Extremophiles" that were adapted to live in extreme environments.

Special lectures and the sessions were as follows:

Opening lecture by Prof. H. W. Jannasch
Session 1: Genetics and Molecular Biology
Session 2: Structure and Function of Proteins
Session 3: Physiology and Metabolisms
Session 4: Ecology and Diversity (including exobiology and subsurface microbiology)
Session 5: Genome
Panel Discussion: Future of Extremophiles
Closing lecture: Prof. Koki Horikoshi

Fig. 11.2 International Congress of Extremophiles at Yokohama, Japan 1998

The congress also included plenary seminars given by invited speakers and several oral presentations selected from the submitted abstracts.

Holger Jannasch of the Woods Hole Oceanographic Institution, USA, gave us the following opening lecture:

The outer edge of our planet's biosphere is determined by organisms that have adapted to life in extreme environmental conditions.

The term "extreme" is based on an average range of conditions for all known forms of life, especially of the eukaryote domain with a slight anthropocentric tinge. Most of the known extremophiles are microbes, of the bacterial and Archaea domains. Their definition comprises adaptations to physiologically highly disparate environmental parameters, such as temperature, pressure, salinity, pH, humidity, toxicity, and low substrate levels.

Microbes adapted to extreme conditions will often not grow at conditions that are "normal" as defined for non-extremophiles.

The larger part of the Earth's biosphere, namely about 80 % by volume, is represented by the oceans' deep and pelagic part. This uniformly well-buffered and moderately saline seawater becomes an extreme environment mainly through its hydrostatic pressure, low temperature (2–4 °C), and low concentrations of energy-yielding dissolved organic substrates for heterotrophic microbial metabolism. Although barophilic and psychrophilic microbes have been studied extensively, research on their uptake and metabolic efficiencies at low levels of organic carbon is more recent.

In view of the fact that the amount of dissolved organic carbon in seawater is in the same order of magnitude as inorganic carbon in the atmosphere, the turnover of the former is important to the release of CO_2 from the ocean. This climate-affecting process may well be controlled by microbial oligotrophic characteristics. The biochemical capabilities of barophilic microbes may differ characteristically from those of non-barophiles. The biotechnological importance in searching for such new barophiles is indicated by JAMSTEC's peerless research efforts in deep-sea microbiology.

The discovery of deep-sea hydrothermal vents provided us with another group of extremophiles: hyperthermophilic Archaea as well as bacteria. The presence of liquid water above 100 °C at deep-sea pressures might be the key for the existence of organisms at the upper, yet unknown, temperature limit of life in general. The fact that many of the isolated extreme thermophiles have been obtained from both deep and shallow marine environments might relate to genetic stability and the microbe's typical dispersibility. In pure culture studies, their maximum growth temperature was recently extended to 113 °C [recently, 122 °C (Takai et al. 2008)]. This limit may well reach higher values with the use of unique source material and novel cultural conditions. High concentrations of heavy metals in hydrothermal fluid and sulfide deposits at deep-sea vent sites pose the scarcely touched research area of microbial adaptations to commonly toxic concentrations of heavy metals.

Fig. 11.3 Holger Jannasch

As implied by the obvious usefulness of many enzymatic properties of extremophiles, much of the driving force in their study comes from their biotechnological potentials. This present international congress is no exception. It covers a broad area as indicated by the large number and various types of microbes that can be defined as extremophiles, and it reflects an increasing interest in their physiological and genetic capabilities and promising biotechnological applications.

Six months after the opening lecture, I received Holgers's last letter, dated 3 August 1998. A short time later I was very sad to learn in an e-mail from Woods Hole that Holger had passed away on 8 September 1998.

The following is extracted from Holger's obituary, which appeared in *Extremophiles* (1999) 3:1–2 (Fig. 11.3).

The extremophile community and microbiology as a whole suffered a grievous loss with the death of Holger Jannasch at his home after a long battle with cancer. The impact of his long career was not restricted to the discipline of microbiology, and he was well known in the wider scientific community as an individual whose scientific work, communication skills, and inspirational qualities transcended subject boundaries.

After completing a doctoral degree in biology from Göttingen University in 1955, Holger held a postdoctoral position at the Scripps Institute of Oceanography with Claude Zobell, one of his early mentors. There he first met C.B. van Niel, whom he was later to call "the scientist of my life." After a further postdoctoral position at the University of Wisconsin, Holger returned briefly to Göttingen as an assistant professor before moving to Woods Hole as a senior scientist in 1963. He was, however, to retain his connection with Göttingen University as Privatdozent there until his death, and he regularly participated in the marine course of the Göttingen Institute, held in the historic Satazione Zoologica in Naples, a place of which he was very fond.

His work at Woods Hole can be divided into three general areas, roughly corresponding to the past three decades. Early seminal work on microbial growth kinetics in seawater using chemostats is perhaps better known to the older ones among us, who as students struggled to cope with the then new mathematical approaches. The loss and subsequent recovery of the submersible *Alvin* spawned a second phase of research into life under high pressures, which defined the technologies for collecting, culturing, and sampling microbial populations from the abyssal depths of the oceans, essentially laying the foundations for the major research programs currently under way in the United States, France and Japan.

The realisation that in these environments chemolithoautotrophic bacteria take over the role of green plants must surely rank as one of the great discoveries of the twentieth century. Holger's work on defining the hydrothermal vent systems and the bacteria that underpin the ecology is now to be found in every microbiology textbook.

Holger served on many editorial boards, including those of the *Journal of Marine Research, Limnology and Oceanology, Applied and Environmental Microbiology, Archives of Microbiology*, and *Marine Biotechnology*, in addition to his association with *Extremophiles*. He was the recipient of many awards and honorary memberships over the years, most recently the Fulbright Distinguished Scholar Award in 1992, the Fellowship of the American Society for Microbiology in 1993, and the rare honour of being elected a Foreign Associate of the National Academy of Science in 1995.

Many graduate students, postdoctoral fellows, and visiting scholars have been mentored by him over the years. His enthusiasm for teaching was communicated in many ways, notably through serving from 1971 to 1980 as Director of the world-renowned Woods Hole Microbial Ecology Course, traditionally maintained and taught by disciples of the Delft school of microbiology. Holger always maintained that he was part of that heritage.

He is remembered as one of the foremost microbiologists of the twentieth century.

An International Conference for Extremophiles has been held every other year since then: in Germany in 2000; Italy 2002; USA 2004; France 2006; South Africa 2008; Portugal 2010; Spain 2012; and Russia 2014. These meetings, of course, have tremendously advanced extremophile microbiology.

To promote this field further, as the launch President and with the collaboration of the managing editors of *Extremophiles*, I set up the International Society for Extremophiles. After the American conference in 2004, I retired as President (2001–2004) and Prof. Garo Antranikian from Germany was elected President (2005–2011), followed by Prof. Helena Santos from Portugal (2011–2014), and Prof. Anronio Ventosa from Spain (2014 until present).

Awarded Honorary Doctorate of Science *Honoris Causa* from University of Kent

On 10 July 2001 I was awarded a doctor's degree at Canterbury Cathedral (Fig. 11.4) and had the opportunity to deliver my thoughts. The last part of my lecture went as follows:

Now, ladies and gentlemen, I would like to point out the quite surprising and bewildering fact that in one gram of your garden soil there are approximately one billion counts of microscopic organisms. Current technology, however, can only isolate a small percentage of these as living organisms in the laboratory, with more than 95 % going completely undetected. This is proof enough that our knowledge is still insufficient. There are an enormous number of unknown living creatures in the earth. It should be our great task to save and conserve them for the benefit of all human beings and the sustainability of our blue planet.

There are no borders in science, but there is a kind of border in a line of thinking derived from different religions and different languages. Even though we may reach the same result, it might be by different ways of thinking. Diversity of cultures and ways of thinking can promote the development of science. Therefore, international collaborations and exchanges of information are an absolute necessity for all investigators.

Fig. 11.4 Honorary Doctorate from University of Kent

Science is the one common language of all human beings. We have just started to communicate with the universe by using science.

Science is just a sheet of white paper. If Michael Faraday wrote on this paper, the paper would become the *Chemical History of a Candle*. If Charles Darwin wrote on the paper, the paper would become *On the Origin of Species*. I am convinced that we will have the opportunity to understand what life is through science.

In an essay, Francis Bacon said: "Travel, in the younger sort, is a part of education; in the elder, a part of experience." To the younger ladies and gentlemen I say: you can learn from different cultures in different countries. Be ambassadors for international understanding and peace. Be ambitious!

Publication of *Extremophiles Handbook*

Studies on extremophiles became more popular after the publication of *Extremophiles*. People began to ask: who discovered extremophiles, such as thermophiles, alkaliphiles, barophiles, etc.? How did they isolate them? However, these very basic questions were not addressed in biology textbooks. Most importantly, what had prompted these discoverers' interest in the first place and how had they begun their research?

For instance, why did Prof. Thomas Brock start works on thermophiles (Fig. 11.5)? His initial interest was in the ecology of Yellowstone National Park in the USA. On 24 October 1984 I met him at the University of Wisconsin, and he told me: "If I had not had an interest in ecology, no one would have thought of thermophiles." It is true. He discovered a form of bacteria in the thermal vents of Yellowstone that can survive at very high temperatures. From these bacteria, enzymes were extracted that are stable at near-boiling temperature. Polymerase chain reaction (PCR), the revolutionary technique for DNA research, depends on Taq polymerase, an enzyme from *Thermus aquaticus* that he first isolated from a hot spring in Yellowstone National Park.

The story of my own discovery is similar. If I had not visited Florence, I would not have discovered alkaliphiles. There is such a tale behind every new field of study.

I decided that I wanted to publish a new handbook on extremophiles and I discussed the idea with Aiko Hiraguchi from Springer in May 2008. I was to be the chief editor, with Alan Bull, Garo Antranikian, Frank Robb, and Karl Stetter as co-editors.

On 12 December 2008, I had a meeting at the Technical University of Hamburg-Harburg with Prof. Antranikian and Dr. Dieter Czeschlik, the Editorial Director, Life Sciences, at Springer. We discussed the draft of the handbook's organisation written earlier and agreed on the following modifications:

Fig. 11.5 Thomas Brock and his wife Kathie at Chicago, IL, in 1996

- The handbook would have approximately 90 chapters altogether.
- The editors would be Koki Horikoshi, Garo Antranikian, Alan Bull, Frank Robb, and Karl Stetter (Fig. 11.6).
- According to our current pricing scheme and the envisaged size of 1,000 pages, the Handbook would have a retail price of €399.

We planned an average size of 15 pages per chapter. However, in the final concept (table of contents) of the book, the editors were asked to fill in the planned individual size for each chapter because that information was to be included in the authors' contracts. The handbook had to be released in time for the next International Extremophiles Congress, 12–16 September 2010 in the Azores. Accordingly, all final manuscripts were to be with Springer by the end of 2009.

After returning to Tokyo, almost every day on the phone and by e-mail I asked friends to write manuscripts for the handbook. Finally the book was published: 1,300 pages with 76 contributors, 13 parts, and 58 chapters: http://link.springer.com/978-4-431-53898-1. Volumes I and II were published in January 2011 by Springer (Figs. 11.7 and 11.8).

Fig. 11.6 The editors of *Extremophiles Handbook*. *From the left*, Koki Horikoshi, Alan Bull, Garo Antranikian, Frank Robb, and Karl Stetter

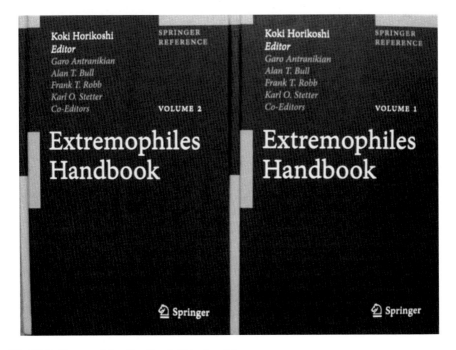

Fig. 11.7 *Extremophiles Handbook*, Vols. 1 and 2

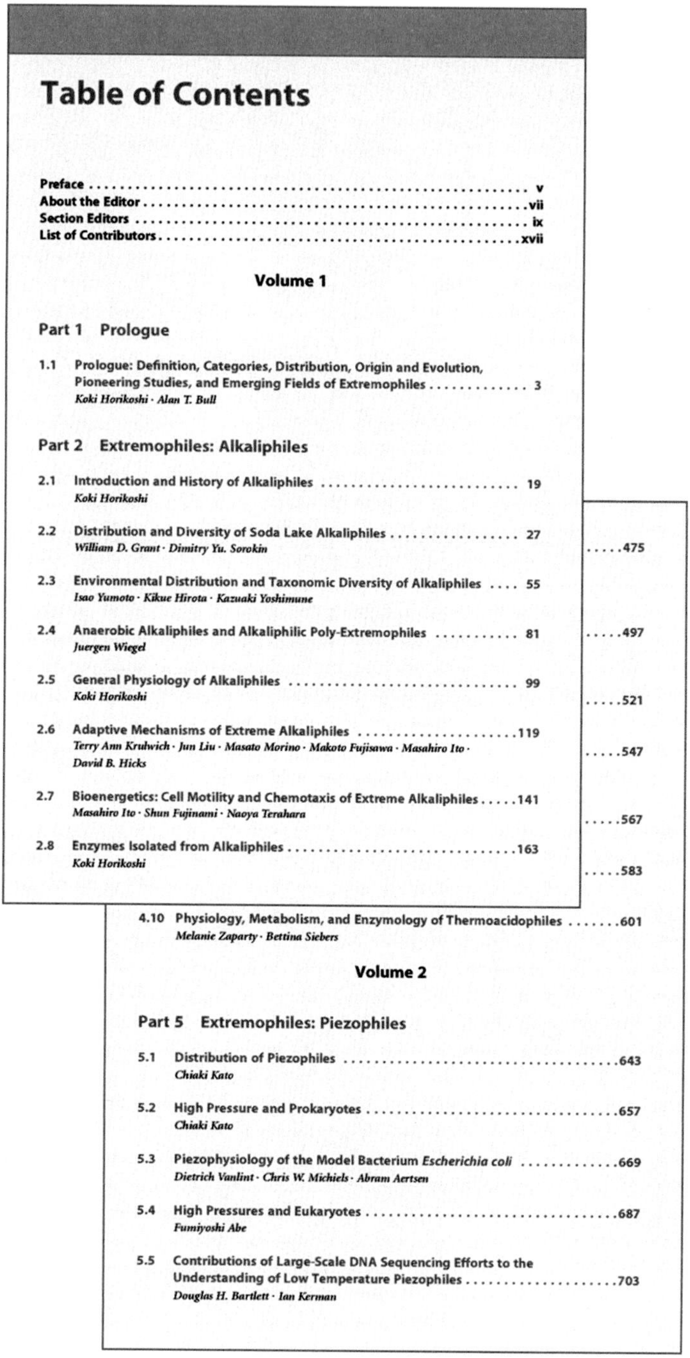

Table of Contents

Fig. 11.8 Table of contents of the *Extremophiles Handbook*

Chapter 12
My Lifelong Friend Hubert Gottschling

When I was putting my papers in order the other day, I happened to find a yellowed airline ticket for Lafayette, Indiana, via San Francisco and Chicago. I studied at Purdue University, in Indiana, as a Fulbright student in 1958. Everything was new to me at that time and I eagerly anticipated the new life ahead of me.

The airport of Purdue University turned out to be quite important in my life. Three weeks after my arrival, I picked up my Major Professor, Henry Koffler, from the airport. He was with a handsome young man whom I did not know. That man was Dr. Hubert Gottschling, who would have a considerable influence on my future. In fact, my encounter with Hubert at the airport was a turning point in my life. Hubert, who had studied medicine at Heidelberg University, came to the United States to conduct research in microbiology. Soon after his arrival, we became good friends, sharing common interests in science, art and philosophy. We exchanged ideas through English in an attempt to understand the different cultures of East and West. I was particularly impressed with his idea that science belongs in the field of philosophy. He himself had attended the Latin lectures of Martin Heidegger and Karl Jaspers at Heidelberg University. He said that not only students of liberal arts but also those who majored in science had attended. From this I learnt that there is no difference in the intellectual hunger of students of science and those of the liberal arts.

I began to be attracted by paintings in the Art Institute of Chicago by such artists as Dürer, El Greco, and Titian, none of which could be seen in Japan. The great works of Titian especially attracted me, and I became determined to see all of his paintings.

After finishing my work on lytic enzymes of *Aspergillus oryzae*, I started a peripatetic journey in search of Titian, with a guidebook annotated in Hubert's own writing. It was the middle of January 1960. Staying at a pension in front of Florence Station, I made frequent visits to the Uffizi Gallery. When I went one evening to the Pitti Palace, where a special exhibition of Titian was being held, I climbed up to the garden in the back. I was greatly impressed and wondered whether I would see this beautiful sight again someday.

© Springer Japan 2016
K. Horikoshi, *Extremophiles*, DOI 10.1007/978-4-431-55408-0_12

Fig. 12.1 Hubert Gottschling

Feeling like an artist myself, I went back to Purdue University at the end of February. I had not imagined that the strong impression in the Pitti Palace would lead to my academic success: the discovery of alkaliphilic microorganisms.

Before long, Hubert moved to the University of Wisconsin and I returned to RIKEN in Japan. We did not see each other for the next 8 years. In spite of my happy family life, I fell into a slump in my research work. Finally, I made up my mind to go abroad again in 1968. In Berlin, at a time when the Cold War was still going on, he met me at Berlin-Tempelhof Airport. Hubert had become an assistant professor at the Berlin Free University. I stayed at his house of Bredschneiderstrasse and we drank beer and talked all night, celebrating our reunion (Fig. 12.1).

I visited the museum in Dahlem and it rekindled my enthusiasm for Titian. I could wait no longer. In November 1968 I returned to Florence and stood once more in the garden of the Pitti Palace, with its great view spread before me. It seemed to me that the city of Florence was a symbol of the Renaissance, containing the past and the present.

It could be said that the Japanese in the Muromachi Era (1392–1573) could never have imagined the existence of such an extraordinary Renaissance happening on the other side of the globe. Then an idea sprang to my mind: there must surely be an undiscovered culture in the world of microbiology. Was there a different type of microbiological group existing under alkaline conditions? I seized the opportunity to pursue the discovery of alkaliphilic microorganism and to establish the microbiology of alkaliphiles.

My research went along smoothly and I visited Hubert's house in Berlin several times a year.

Hubert married Gisela, a psychiatrist, and they had a son, Marc (Fig. 12.2). They visited us in Tokyo many times and stayed in our house. We had many excellent opportunities to understand the cultures of our two countries. And it was he who urged me to adopt the term "extremophiles" for microorganisms that love extreme environments.

Fig. 12.2 Koki with the Hubert family in Berlin: Gisela, Marc, and Hubert

Until 1989 armed East German border guards watched out for any East Germans trying to cross the border, and many were killed by the guards. Furthermore, in West Berlin, there was the Soviet War Memorial in the Tiergarten. Two of the first Soviet tanks to enter Berlin at the end of the Second World War were installed there and protected by armed guards. Of course, this enclave belonged to the Soviet Union, and was fenced off with barbed wire. When Toshiaki, Sachiko and I visited the memorial with Hubert in 1977 (Fig. 12.3), we were confronted with a forbidding notice: "Entry prohibited" in English and "Eintritt Verboten" in German. "If the Soviet soldiers shouldered, you should escape into the barricades as soon as possible." The English, presumably translated from German or Russian, might have been somewhat erratic but the threat was clear.

Sachiko and I had the opportunity to visit numerous museums, including those in East Berlin, under Hubert's guidance. If he had not taken us to East Berlin, we would not have been able to visit the Humboldt-Universität zu Berlin, the Pergamon Museum, etc., during the Cold War.

On 10 November 1989, we saw the fall of the Berlin Wall on TV in Japan. I recognised Hubert's son Marc among the crowd tearing it down. I phoned Hubert immediately, and he confirmed that Marc was on the wall. What a small world we live in.

In May 1990, we spent a week at Hubert's home near the wall. We chipped away at the wall ourselves and took a trip into East Germany (Fig. 12.4). Nobody asked for our passports, and we were able to walk freely and to visit the churches. Moreover, we were able to cross the Glienicker Bridge that had been used only for the exchange of spies during the Cold War, and we went to Potsdam. I brought home a piece of the wall, an insulator for high-voltage wire, and some empty cartridges as souvenirs.

Fig. 12.3 Soviet War Memorial in Tiergarten, Berlin

Fig. 12.4 A piece of the
wall, insulator for high-
voltage wire, and empty
cartridges

The wall itself was largely destroyed in 1990, and its fall paved the way for German reunification, which was formally concluded on 3 October 1990. Peace at last.

As we were leaving Berlin, Hubert said, "I feel lonely, and am not feeling well." I told him: "We will be back in the autumn."

A month after our return to Tokyo, I received an airmail from Gisela. My lifelong friend Hubert had died of a heart attack. We phoned Gisela to offer our condolences.

Chapter 13
Epilogue

I do not know what I may appear to the world; but to myself I seem to have been only like a boy playing on the seashore, and diverting myself in now and then finding a smoother pebble or a prettier shell than ordinary, whilst the great ocean of truth lay all undiscovered before me. Isaac Newton. (From Brewster, *Memoirs of Newton* (1855))

Progress in microbiology requires well-trained microbiologists who think "outside the box" and who will accept new challenges and investigate them with creative energy. The task facing academic institutions is to lead the way in training the microbiologists of the future. These microbiologists will have to be versed not only in the basics of microbiology, but also in molecular biology, biochemistry, ecology, chemistry, computer science, physics, and mathematics. They must be able to improvise, to use their knowledge and imagination, and to develop a higher level of understanding of microbial life.

The study of communities of microorganisms, the characterisation of many unknown bacteria in the ocean and soil, the use of new isolates instead of laboratory strains, and the further analysis of "uncultivatable" organisms present an unlimited future for those who are interested in studying bacteria.

Fifty or so years ago few microbiologists studied microorganisms living in extreme environments. Almost every microbiologist had thought of these environments as abnormal, and that no living creatures could survive there. No one thought there were a host of microorganisms that love extreme environments. And no microbiologist could have expected to see such a thing as the *Extremophiles Handbook*.

These days we cannot discuss biology, especially microbiology, without knowledge of extremophiles. As the *Extremophiles Handbook* makes clear, extremophiles have been found all over the Earth, and they are no longer exotic.

The question arises of what we do now. The answer is that nobody knows. If we knew the way, we would already have taken it.

The "New Frontiers" section of the *Extremophiles Handbook* should illustrate just one part of a future field that is bound to develop very quickly and ever more broadly. Among other aspects, gravity, water activity, etc., in extreme environments have not been fully studied. These extreme environments would be compatible with

© Springer Japan 2016
K. Horikoshi, *Extremophiles*, DOI 10.1007/978-4-431-55408-0_13

other factors, some of which, of course, are already referenced in the *Extremophiles Handbook*.

There is an interesting story in *Nature* about Low Life (Mascarelli 2009). In the 1980s and 1990s, some of the first missions of the Ocean Drilling Program made it possible for researchers to dig deeper than ever before. When, in 1990, Professor John Parkes of Cardiff University and his colleagues submitted to *Nature* results that showed that bacteria could colonise much greater subsurface depths than previously thought, they were met with "very sceptical reviews" and the paper was rejected. But in 1994, the team finally succeeded in publishing their results and reported viable microbes living in ocean sediments at depths greater than 500 m below the seabed.

In another example, Professor Akira Inoue of Toyo University discovered toluene-tolerant microorganisms (*Pseudomonas putida* IH2000) in forest soil from Kyushu, Japan. His follow-up experiments produced a further surprise: he found such microorganisms in many places around the world, including in seawater. He submitted his data to *Nature*, but the referee was initially sceptical. Although the manuscript showed the growth curves of the bacteria in the presence or absence of toluene, one of the reviewers did not trust Inoue's experiments. This was no wonder because toluene is very toxic. Finally, Inoue sent a letter containing photographs showing very good growth in the culture test tubes containing 30 % toluene. If the reviewer had further questions, he added, please come to Japan, all expenses paid. Fortunately the reviewers believed Inoue's data, and his discovery was published in *Nature* in 1989.

Both Parkes and Inoue, of course, have contributed to the *Extremophiles Handbook*.

Once such results have been introduced in textbooks and become common knowledge, the names of the discoverers gradually fade away.

As concerns the deep biosphere, the *Extremophiles Handbook* demonstrates that there are many microorganisms in the deep-sea sediments and crust. The close collaboration with Prof. Antranikian's group from Germany has shown that in addition to the broad diversity of microorganisms in the deep sea, a huge number of enzymes can be identified that are of interest for diverse industrial applications (Fig. 13.1). Dramatic progress in cultivation methods has made it clear that these microorganisms are entirely different from *Escherichia coli* cultivated in the presence of organic compounds because the nutrients of those organisms are synthesised by using solar energy.

In contrast, deep-sea microorganism communities graze on deeply buried organic carbon, such as methane. The generation times of sub-seafloor microorganism are, of course, not clear, but deep terrestrial life forms in South African gold mines are estimated to reproduce once every 1,000 years. How can we measure such a long generation time in our labs? However, some might say that it is not such a long time on the geological timescale.

According to Professor P.C.W. Davies of Arizona State University, our planet may have forms of "shadow life" that are different from life as we know it (Morgan 2009). This "shadow life" may be hidden in various extreme environments such as

Fig. 13.1 Prof. Garo Antranikian and Prof. Koki Horikoshi on the deck of the drilling vessel *Chikyu* in 2009

arsenic lakes, deserts, deep seas, and hydrothermal vents. Shadow life could even be living among us in forms that we do not yet recognise. In addition, Davies posits an intriguing idea: our microscopes are customised for life as we know it; and this is why we have not found shadow life with its different biochemistry.

The signatures of an ecologically integrated shadow biosphere might be revealed as a distinctively different organic chemistry, such as use of a wider range of amino acids, homochirality, the use of a novel source of energy (e.g., UV radiation, higher pressure, higher gravity), or the use of a selection of elements different from those used by standard life (Davies et al. 2009). This shadow life may be hidden in various natural and man-made environments such as hyper-arid deserts, hydrothermal vents, or deep subsurfaces (Fig. 13.2), and those environments exposed to intense radiation, salinity, or pH, or depleted in key standard life elements. A recent study shows new constraints on the size and distribution of microbial biomass beneath the ocean floors that remain unclear (Hinrichs and Inagaki 2012). To understand shadow science (physics, chemistry, biochemistry, biology, etc.), we have to be virtual scientists. Otherwise we shall never understand it in our conventional wet labs.

For example, we do not use the concept of "time axis" in biology, including biochemistry and genetics, so it is difficult to deduce predominant metabolic pathways using computers. We cannot estimate the generation times of extremophiles, so, naturally, we cannot speculate on the generation times of shadow life. They have individual biological clocks, so the scale of their time axis will be different.

Fig. 13.2 Drilling vessel *Chikyu*

If we upload our life data onto computers running our operating systems and software, the computers can produce reasonable answers. However, no such computer can give us any information from data of shadow life, including from the early earth or other planets. An entirely different operating system is required for shadow life. It is as yet premature for us to extend our thinking to the field of astrobiology.

We are convinced that microbiologists can find an opportunity to create a new microbiology, if they only know how to ask the extremophiles.

Short Story

Serendipitous Discoveries in Graduate School

In an old notebook I found an account of why I selected Prof Sakaguchi's microbiology laboratory. In 1955 I had built an HF oscillator to measure the dielectric constants of an alcohol and water mixture in his lab. Prof Sakaguchi used to say: "Microorganisms can do anything you want." His lab was actually a gathering place for ambitious people and during that time I met many great scientists there. General microbiology, industrial microbiology, radiation biology, sake brewing, physics of alcohol and water, etc. were major research themes. About ten researchers, including me, formed a pirate group whose enjoyable philosophy was "Debate = Argument + Sake drinking – Sakaguchi."

In 1952 I entered the University of Tokyo to study mathematics, physics, chemistry, etc., for 4 years. In 1956 I joined the Chemistry Department of the university as a graduate student.

As I explain in Chap. 3, in 1956 Prof Sakaguchi gave me the research theme for my doctoral thesis, "Autolysis of *Aspergillus oryzae*." He believed that it was an autolysate of *Aspergillus oryzae* that gave sake its flavour. Every day I cultured 50 stock strains of *Asp. oryzae*. I was required to taste the cultured fluid by *bero*-meter. I hated this theme.

Isolation of Fungi-Lysing Bacteria

One day in November I found one cultivation flask in which the mycelia of *Aspergillus oryzae* had completely disappeared. The previous night when I had inspected the flasks, the mould had been flourishing in all the flasks. I remembered the spectacular pictures I had seen of how bacteria thrived and moved. Yet no mycelium could be seen under the microscope. I showed the flask and asked fellow

© Springer Japan 2016
K. Horikoshi, *Extremophiles*, DOI 10.1007/978-4-431-55408-0

members of the pirate group what they thought had happened. Masahiro Takahashi (the leader of the pirate group) and Tsuneo Yasui said that my poor cultivation technique had resulted in the bacteria becoming contaminated. Hideo Nakamura and Kazuo Izaki suggested that the contaminated bacteria should lyse *Aspergillus oryzae*. This might be the opposite of Fleming's discovery of penicillin. These discussions led me to future serendipitous research work.

At the beginning of 1957, Dr. Mikio Amaha, who had been at the University of Illinois on a Fulbright post-doctorate scholarship, returned to the Sakaguchi Lab. I talked to him about my research. He found my work very interesting, and suggested that we go to explain the details to Prof. Sakaguchi. As a result, and thanks to Mikio, my subject was changed from "Autolysis of *Aspergillus oryzae*" to "Lysis of *Aspergillus oryzae*."

In autumn 1957 I was keen to report my data so I asked Mikio which journal he thought would be the best. He strongly recommended *Nature,* which greatly surprised me. As far as I knew, no research work by our department had ever been published in *Nature*.

I wrote a draft of "Lysis of fungal mycelia by bacterial enzymes," which was revised and polished for *Nature* by Mikio. Although I posted it to *Nature* by air mail, I received no answer.

At the beginning of 1958, I received a galley of our paper, reprints from *Nature*, and several hundreds letters of request on the same day. All of them had been sent by sea mail from the USA and Europe. I could not understand what had happened.

Nevertheless, my first report was published in *Nature* on March 29, 1958 (Horikoshi and Iida 1958).

That first paper changed my life. One of the letters was from Professor Henry Koffler, of Purdue University, Indiana. He wished to use my *Asp. oryzae*-lysing *Bacillus circulans* to study the cell-surface structure of *Penicillium chrysogenum*.

I showed Koffler's letter to Prof. Sakaguchi to get his permission. He told me I should go to Koffler's lab and join his research. I wrote to Koffler about my professor's consent and Koffler kindly invited me to Purdue and even arranged a scholarship for me. I accepted his invitation and decided to travel to Lafayette, Indiana. Fortunately, Fulbright Funding sponsored my travel expenses.

Mikio had one further suggestion for me: in those days Japan had only one ship, the *Hikawa-maru,* carrying passengers across the Pacific Ocean. The crossing would take 3 to 4 weeks, with another 3 or 4 days by train to Lafayette. So, he recommended that I go by air.

Fortunately I had Koffler's letter in which he said that he and his colleagues wanted to use my bacteria from the middle of August.

I went to the Fulbright office to ask if I could go by aeroplane as the *Hikawamaru* was not scheduled to leave Japan until the end of August, too late for me. The officer kindly exchanged my ticket for a JAL flight from Tokyo to Lafayette on 8 August.

In the evening of 8 August, members of the pirate group, Prof. Iida of RIKEN, my mother, and relatives came to see me off at Haneda Airport. The DC-7 aeroplane could only take 15 passengers. The next morning the captain came to my seat and explained that we had just crossed the International Date Line, so that we had begun 8 August over again. He gave me a certificate showing that I had passed over the date line.

On 15 August I arrived at West Lafayette via San Francisco and the Davis campus of the University of California. Prof. Koffler, however, was in Switzerland, and the lab was in the process of moving from Stanley Hall to the brand-new Life Sciences Building. All I could do was wait for his return.

Three weeks later I met Henry Koffler and his wife Phyllis at the university airport. Also there was a man named Hubert Gottschling, from the University of Heidelberg, West Germany. Before long we became close friends and it was he who opened European culture to me. If I had not met him, I might not have visited Florence in 1968. In all probability, I would not have discovered alkaliphilic microorganisms.

The first thing I had to find was the substrate specificity of mould-lysing enzyme(s), but the purified enzyme could not hydrolyse any of the available poly-saccharides. In March 1959 John Greenawalt, working in the same room as me, gave me a polysaccharide, laminarin (1,3-β-glucan isolated from seaweed), that he had obtained from a friend in Norway. The enzyme hydrolysed the polysaccharide, and laminaribiose was detected in the hydrolyzate. This result indicated that the mould-lysing enzyme was endo-1,3-β-glucanase. I was delighted to have found the substrate specificity. The next morning I told John about the previous night's results and I still remember his warm congratulations.

I had a half-time research assistant scholarship of 150 dollars a month. Henry kindly changed this from a half-time to a full-time scholarship and increased the stipend to 300 dollars a month, which allowed me to save about 150 dollars a month. I conceived a wish to visit Europe in January 1960, using this extra money. Hubert recommended a number of places to visit in Europe, including Germany and Italy.

Launching a New Journal, *Extremophiles: Microbial Life Under Extreme Conditions*

At the end of the 1980s, extremophiles were not popular with conventional micro-biologists. Some sneered that only abnormal microbiologists would study microbi-ology in abnormal environments, although many extremophile enzymes were being used in industrial applications. No journals would willingly accept reports of these extremophiles, and our work had been rejected many times.

In June 1996, EU researchers decided to hold the First International Meeting on Extremophiles in Estoril, Spain. I was never content to be second, so the best thing I could do was to establish a new society and journal for extremophiles. I talked with Ms. Aiko Hiraguchi, a publishing editor at Springer Japan, and we vaguely agreed, if possible, to publish an entirely new journal. We planned to attend the meeting in Estoril and hold discussions with EU scientists. How many researchers would embrace the launch of a new journal for extremophiles? Clearly, we needed their agreement, otherwise we could not proceed.

At that time e-mail was not common, so I sent airmail letters to various extremophile researchers. They agreed to cooperate with me in publishing a journal about extremophiles, but among them they had scant experience in founding a new journal.

As the chief editor, I selected the following five eminent microbiologists as managing editors: William D. Grant (University of Leicester, UK), Garo Antranikian (Technical University of Hamburg-Harburg, Germany), Karl Stetter (University of Regensburg, Germany), Terry A. Krulwich (Mount Sinai School of Medicine, New York, USA), and Juergen Wiegel (University of Georgia, Athens, USA). I visited their offices or houses to talk to them face to face. In addition, I phoned more than 30 researchers and 20 biologists, who agreed to join the journal as editorial board members.

In May 1996 I attended the American Society of Microbiology meeting in New Orleans. There I had lunch with Dr. Dieter Czeschlik (Editorial Director, Life Sciences, Springer-Verlag in Heidelberg), who also agreed to publish the new journal.

At the First International Meeting of Extremophiles in Estoril, Spain, I mentioned that I would be happy to host the Second International Meeting of Extremophiles in Yokohama, Japan. And then I talked about the new journal *Extremophiles*. The journal, I said, would be published in 1997, and my office was awaiting their manuscripts.

In February 1997, *Extremophiles* was published and the first issue set out its aims as follows:

Extremophiles publishes descriptions of all aspects of research as well as research of a genetic and molecular nature that focuses on topics of practical value and basic research on microorganisms under extreme conditions.

Extremophiles is an international journal featuring articles of original research and mini-review articles on those areas of microorganisms that thrive in extreme conditions, including bacteria, yeasts, fungi, protozoa, viruses, and other simple eucaryotic organisms.

The Editors invite submissions in the areas of:

Isolation, growth, development, and morphogenesis,
Ultrastructure,
Biochemistry,
Physiology,
Metabolism,
Molecular and cell biology,
Biotechnology,
Symbiosis,
Ecological and environmental physiology,
Microbial processes associated with environmental pollution,
Current methodological development.

The first issue contained five original papers and one review paper. I was very happy to have the new journal.

In 2008, Garo Antranikian took over as Chief Editor, and I became the Founding Editor.

International Society for Extremophiles

The International Society for Extremophiles (ISE) was founded on the initiative of the members of the editorial board of *Extremophiles* in 2002. The ISE is intended to be a forum for the exchange of information and experience in the rapidly growing field of research on extremophiles. The main objectives of the society are:

- Support for scientists who dedicate their research to the investigation of extremophilic microorganisms and their enzymes for basic and applied research.
- Creating benefits for the ISE members, including reduced subscriptions for the *Extremophiles* journal, which is the official journal of the ISE.
- Support for local organisers of the biennial Extremophiles Conferences.
- Scholarships and travel grants for young scientists.

As the former chief editor of *Extremophiles*, I became the founding president of the ISE. I was supported by Prof. Antranikian (President from 2004 to 2010 and chief editor of *Extremophiles*). During the Extremophiles Conference in Baltimore, 2004, Prof. Antranikian took over the position from me and the ISE organisation was transferred to Hamburg, Germany.

For "outstanding contributions to the research of extremophiles and commitment to the International Society for Extremophiles" I was elected lifetime honorary president of the society.

Prof. Helena Santos (New University of Lisbon) became the new President of the ISE in January 2011 (elected during the Extremophiles Conference 2010 in the Azores).

New Department of Extreme Biology

At the beginning of 1992 I received a phone call from the chairperson of the board of directors of Toyo University, Dr. Masajyuro Shiokawa. He wanted to meet me in his office as soon as possible. I visited him, whereupon he invited me to Toyo University as a full professor to create a biology department in his university. I asked him why he wanted a new biology department. His answer impressed me deeply. "Biology," he said, "should be one of the most important fields in Japan."

I had created many research laboratories – at RIKEN, the ERATO Superbugs Project, the Tokyo Institute of Technology, and JAMSTEC. However, I had never established a new department, even if in a private university. I asked him what type of biology department he had in mind, and explained that, if possible, I would like to set up an extremophile department. Dr Shiokawa told me that I could do anything I wanted. I was quick to accept his proposal.

In April 1993, I moved temporarily to the Department of Engineering and collected the CVs of candidates to join me in the new department. Not only biologists, but also physicists, chemists, and information scientists were the preferred candidates. I personally selected the professors from more than 100 CVs.

In October 1994, I had a meeting with the Ministry of Education. At that time, I was in London at an international meeting, but I returned to Tokyo for 1 day only to attend the interview. They told me that my extremophile work was unique and very useful in industrial applications. The whole thing took less than 10 min! The next day I went back to London at my own expense.

I found a number of differences between a humanities department and a biology department. For example, the clerical staff asked me how many days I would be in the study rooms. I told them almost 7 days a week, because we were going to cultivate living things.

By April 1996, we had 189 students from 3,100 candidates, and after 5 years a number of students received doctorate degrees in the field of extremophiles.

Timeline of Koki Horikoshi's Life

1932 October 28	Born in Saitama Prefecture, Japan.
1939	Entered Nakatsu Daisan Primary School in Osaka.
1940	Moved to Ichikawa Primary School in Chiba Prefecture.
1941 December 8	The Pacific War broke out.
1942 April 18	The first air raid on Tokyo area by JAMS Harold Doolittle.
1943 April	A fateful encounter with Mr. Kazuo Hori who inculcated in me a deeper interest in science.
1945 March	Evacuated to Kumagaya, Saitama Prefecture.
1945 August 14	Kumagaya area firebombed by Boeing B-29 bombers. The war ended on 15 August.
1946 February 20	Father, Tomozou, passed away and our family moved to Hanyu, Saitama Prefecture. Changed to Fudoka High School.
1952 April	Entered the University of Tokyo to study mathematics, physics, chemistry, etc., for 4 years.
1956 April	Joined the Chemistry Department of the University as a graduate student. Major professor was Kin-ichiro Sakaguchi.
1956 October 23	Discovered mould-lysing bacterium *Bacillus circulans* IAM 1165.
1957 November 5	Sent the manuscript to *Nature*.
1958 March 29	Horikoshi K, Iida S (1958) "Lysis of fungal mycelia by bacterial enzymes" published in Nature 181:917–918.
1958 August 8	Travelled to Purdue University in the USA to study enzymology as a Fulbright student. Met a lifelong friend, Hubert Gottschling, there.
1959 January 17	Horikoshi K, Iida S (1959) "Effect of lytic enzyme from *Bacillus circulans* and chitinase from *Streptomyces* sp. on *Aspergillus oryzae*" published in Nature 183:186–187.

© Springer Japan 2016
K. Horikoshi, *Extremophiles*, DOI 10.1007/978-4-431-55408-0

1960 January 7	Fifty days' holiday to visit Europe.
1960 June 8	Return to Japan and the University of Tokyo. The major professor was Kei Arima, and Koki continued to study the fine structure of cell walls of *Aspergillus oryzae*.
1963 March 31	Received doctor's degree.
1963 April	On research staff of RIKEN Institute.
1963 November 14	Married Sachiko Hamada.
1994 September 29	Son Toshiaki born.
1996 April	Awarded prize of the Society of Agricultural Chemistry for mould-lysing enzymes.
	Moved to Prof. Roy Doi's laboratory at the Davis campus of the University of California as a post doctorate to study the protein synthesis of *Bacillus subtilis*.
1998 June	Returned to RIKEN Institute.
1967 October	Flew to Europe, in the hope that it would lift me out of the slump. At the end of October, visited Florence, Italy, and saw the Renaissance buildings, a style of architecture utterly different from that of Japan. Then suddenly, heard a voice whispering in my ears, "There could be a whole new world of microorganisms in different unexplored cultures." Upon return to Japan on 2 November, I prepared two alkaline media containing 1 % sodium carbonate "Horikoshi-I and Horikoshi-II," put small amounts of soil collected from various areas within RIKEN Institute into 30 test tubes, and incubated them overnight at 37 °C. To my surprise, various microorganisms flourished in all 30 test tubes. Isolated a great number of alkaliphilic microorganisms and purified many alkaline enzymes.
	Here was a new alkaline world that was utterly different from the neutral world. Named these microorganisms, which thrive in alkaline environments, "alkaliphiles."
1969	Alkaline proteases, alkaline amylases, and alkaline pectinases isolated from alkaliphiles.
1970	Alkaline amylase 38–2 CGTase discovered.
September	Promoted to a sub-chairman of the laboratory.
1971	Enzymes produced by alkaliphiles were published in *Journal of Agricultural Chemistry*. There were, however, no responses to the research work.
1972	Isolated alkaliphilic *Aeromonas* sp. no. 212 producing a cellulase that hydrolysed cellulose in human faeces. The Japanese economy, however, slowly began to grow. Flush toilets became popular and the classic toilet disappeared, so no plant was ever constructed.
1973	Alkaliphilic bacteria found to require sodium ion for their growth.
	Industrial production of cyclodextrin started.

1974 May	Promoted to the head of research laboratory. Cyclodextrin became commercially available.
1976	*Bacillus* sp. 38–2 CGTase was purified.
1977	Discovery that sodium ion is needed to incorporate nutrients into alkaliphile cells.
1978	Intracellular enzyme systems including protein-synthesizing systems found to be almost the same as those of *Bacillus subtilis*.
1979	Award from the Ministry of Sciences.
1980	Ichimura Prize.
1981	Inoue Harunari Prize.
1982	Ohokochi Prize.
	Horikoshi K, Akiba T (1982) Alkalophilic microorganisms: a new microbial world. Springer, Heidelberg.
October	Inspection tour: University and Research Institutions in UK. Met Kazuko (Peter Ingham's wife) as interpreter.
1983	Worked on fine structure of cell walls of alkaliphilic *Bacillus* sp.
1984	Superbugs project in ERATO launched.
1985	Novel triangular archaeon *Haloarcula japonica* isolated from the Noto Peninsula.
1987	Discovered organic solvent-tolerant microorganisms.
April 1	New laundry detergent, "Attack", containing alkaline cellulase as an additive launched in Japan by Kao Company.
	Awarded the Medal with Purple Ribbon from Japanese Government.
1988–1994	Professor of Tokyo Institute of Technology and head of research laboratory of RIKEN Institute.
	Ichimura Prize in industrial field.
1989	Inoue A, Horikoshi K (1989) "A *Pseudomonas* thrives in high concentrations of toluene" accepted for publication in Nature 338:264–266.
	Awarded Umetaro Suzuki Prize of The Society of Agricultural Chemistry for work on alkaliphiles.
1990	Discovered pALK fragment which was related to alkaliphily.
1991	The pALK fragment found to be Na^+/H^+ antiporters.
March 11	Presented with the gold medal of the International Institute of Biotechnology by Prince Michael of Kent at the Royal Society, London.
October	Professor Kei Arima Prize.
	Horikoshi K, Grant WD (eds) (1991) Superbugs. Springer, Heidelberg/Tokyo.
	Horikoshi K (1991) Microorganisms in alkaline environments. Kodansha-VCH, Tokyo.

1992	Discovered pALK fragment had 4–5 Na^+/H^+ antiporter.
1993	Acidic polymers in the cell wall found to maintain pH homeostasis.
March	Retired as Professor of Tokyo Institute of Technology and head of research laboratory of RIKEN Institute.
	Granted the title of Professor Emeritus of Tokyo Institute of Technology and Researcher Emeritus of RIKEN Institute.
April	Professor of Toyo University.
	Honda Prize.
1996 July 2	Lecture to the Emperor about the flora on the deepest sea-floor samples from 10,898 m beneath the Pacific Ocean.
	Honorary citizen of Hanyu City.
1997 February	Founded academic journal *Extremophiles* from Springer Tokyo/Heidelberg. Appointed Chief Editor.
1998 January 18–22	Organized 2nd International Congress on Extremophiles in Yokohama.
	Horikoshi K, Grant WD (eds) (1999) Extremophiles. Wiley, New York.
1999	President of Japan Extremophiles Society.
	Horikoshi K (1999) Alkaliphiles. Harwood Academic/ Kodansha, Amsterdam/Tokyo.
	Horikoshi K, Tsujii K (eds) (1999) Extremophiles in deep-sea environments. Springer, Tokyo/Heidelberg.
2000	Complete genome sequence of *Bacillus halodurans* C-125 determined.
2001 July 10	Awarded Honorary Doctorate of Science honoris causa from University of Kent.
	Organized International Extremophiles Society (ISE) and elected as Chairperson.
2002	Awarded The Order of the Rising Sun, Gold Rays with Neck Ribbon.
2003 March	Retired from Toyo University.
2004 April	Director of Deep Star, JAMSTEC.
	Honorary Chairperson of International Extremophiles Society.
2006 July 3	Received the highest prize, "The Japan Academy Prize," for research work on alkaliphiles.
2006	Horikoshi K (2006) Alkaliphiles: genetic properties and applications of enzymes. Springer/Kodansha, Tokyo/ Heidelberg.
September 20	ISE Award 2006 (Brest, France).
2007	Honorary member of Japan Chemical Society.
2008 January	Founding Editor of *Extremophiles*.

December 12	Discussed the draft of *Extremophiles Handbook* with Prof. Garo Antranikian and Dieter Czeschlik of Springer and agreed to publish it.
2010 September 24	ISE Award for Lifetime Achievement (Azores, Portugal).
2011 January	Horikoshi K (ed) (2011) *Extremophiles Handbook*, vols 1 and 2. Springer, Tokyo/Heidelberg.

References

Abe F, Horikoshi K (1998) Analysis of intracellular pH in the yeast *Saccharomyces cerevisiae* under elevated hydrostatic pressure: a study in baro- (piezo-)physiology. Extremophiles 2:223–228

Brock TD (1997) The value of basic research: discovery of *Thermus aquaticus* and other extreme thermophiles. Genetics 146:1207–1210

Brock TD, Freeze H (1969) *Thermus aquaticus* gen. n. and sp. n., a nonsporulating extreme thermophile. J Bacteriol 98:289–297

Davies PCW, Benner SA, Cleland CE, Lineweaver CH, McKay CP, Wolf-Simon F (2009) Signatures of a shadow biosphere. Astrobiology 9:241–249

Deguchi S, Tsudome M, Mukai S, Corkery RW, Ito S, Horikoshi K (2011) Microbial growth at hyperaccelerations up to 403,627 × *g*. Proc Natl Acad Sci USA 108:7997–8002

Doukyu N, Toyoda K, Aono R (2003) Indigo production by *Escherichia coli* carrying phenol hydroxylase gene from *Acinetobacter* sp. strain ST-550 in a water–organic solvent two-phase system. Appl Microbiol Biotechnol 60:720–725

Harada K, Kameda M, Suzuki M, Mitsuhas S (1963) Drug resistance of enteric bacteria. II. Transduction of transferable drug-resistance (R9) factors with phage epsilon. J Bacteriol 86:1332–1963

Hinrichs K, Inagaki F (2012) Downsizing the deep biosphere. Science 338:204–205

Honda H, Kudo T, Ikura Y, Horikoshi K (1985) Two types of xylanases of alkalophilic *Bacillus* sp. No.C-125. Can J Microbiol 31:538–542

Horikoshi K (1971a) Studies on the conidia of *Aspergillus oryzae*. XI. Latent ribonuclease in the conidia of *Aspergillus oryzae*. Biochim Biophys Acta 240:532–540

Horikoshi K (1971b) Production of alkaline enzymes by alkalophilic microorganisms. Part I. Alkaline protease produced by *Bacillus* no. 221. Agric Biol Chem 35:1407–1414

Horikoshi K (1971c) Production of alkaline enzymes by alkalophilic microorganisms. Part II. Alkaline amylase produced by *Bacillus* no. A-40-2. Agric Biol Chem 35:1783–1791

Horikoshi K (1991) Microorganisms in alkaline environments. Kodansha-VCH, Tokyo

Horikoshi K (1999a) Alkaliphiles. Harwood Academic/Kodansha, Amsterdam/Tokyo

Horikoshi K (1999b) Alkaliphiles, some applications of their products for biotechnology. Microbiol Mol Biol Rev 63:735–750

Horikoshi K (2006) Alkaliphiles – genetic properties and applications of enzymes. Kodansha/Springer, Tokyo/Heidelberg

Horikoshi K (ed) (2011) Extremophiles handbook, vol 1 and 2. Springer, Tokyo/Heidelberg

Horikoshi K, Akiba T (1982) Alkalophilic microorganisms: a new microbial world. Springer/Gakkai-Shuppan Center, Heidelberg

Horikoshi K, Doi RH (1968) The NH$_2$-terminal residues of *Bacillus subtilis* proteins. J Biol Chem 243:2381–2384

Horikoshi K, Grant WD (eds) (1998) Extremophiles: microbial life in extreme environments. Wiley-Liss, New York

Horikoshi K, Iida S (1958) Lysis of fungal mycelia by bacterial enzymes. Nature (Lond) 181:917–918

Horikoshi K, Iida S (1959) Effect of lytic enzymes from *Bacillus circulans* and chitinase from *Streptomyces* sp. on *Aspergillus oryzae*. Nature (Lond) 183:186–187

Horikoshi K, Iida S (1964) Studies of the spore coats of fungi. 1. Isolation and composition of the spore coats of *Aspergillus oryzae*. Biochim Biophys Acta 83:197–203

Horikoshi K, Ikeda Y (1965) Trehalase in conidia of *Aspergillus oryzae*. J Bacteriol 91:1883–1887

Horikoshi K, Ikeda Y (1969) Studies on the conidia of *Aspergillus oryzae*. IX. Protein synthesizing activity of dormant conidia. Biochim Biophys Acta 190:187–192

Horikoshi K, Iida S, Ikeda Y (1965) Mannitol and mannitol dehydrogenases in conidia of *Aspergillus oryzae*. J Bacteriol 89:326–330

Inoue A, Horikoshi K (1989) A *Pseudomonas* thrives in high concentrations of toluene. Nature (Lond) 338:264–266

Inoue A, Horikoshi K (1991) *Pseudomonas putida* which can grow in the presence of toluene. Appl Environ Microbiol 57:1560–1562

Kato C, Li L, Tamaoka J, Horikoshi K (1997) Molecular analyses of the sediment of the 11000 m deep Mariana Trench. Extremophiles 1:117–123

Kato C, Li L, Nogi Y, Nakamura Y, Tamaoka J, Horikoshi K (1998) Extremely barophilic bacteria isolated from the Mariana Trench, Challenger Deep, at a depth of 11,000 meters. Appl Environ Microbiol 64:1510–1513

Kinoshita S, Udaka S, Shimono M (1957) Studies of amino acid fermentation. J Gen Appl Microbiol 3:193–199

MacElroy RD (1974) Some comments on the evolution of extremophiles. Biosystems 6:74–75

Mascarelli AL (2009) Low life. Nature (Lond) 459:770–773

Morgan J (2009) Alien life 'may exist among us.' BBC, London. http://news.bbc.co.uk/2/hi/science/nature/7893414.stm

Okazaki W, Akiba T, Horikoshi K, Akahoshi R (1984) Production and properties of two types of xylanases from alkalophilic thermophilic *Bacillus* sp. Appl Microbiol Biotechnol 19:335–340

Okazaki W, Akiba T, Horikoshi K, Akahoshi R (1985) Purification and characterization of xylanases from alkalophilic thermophilic *Bacillus* spp. Agric Biol Chem 49:2033–2039

Takai K, Nakaura K, Toki T, Tsunogao U, Miyazaki M, Miyazaki J, Hirayama H, Nakgawa S, Nunoura T, Horikoshi K (2008) Cell proliferation at 122 °C and isotopically heavy CH_4 production by a hyperthermophilic methanogen under high-pressure cultivation. Proc Natl Acad Sci USA 105:10949–10954

Takami H, Inoue A, Fujii F, Horikoshi K (1997) Microbial flora in the deepest sea mud of the Mariana Trench. FEMS Microbiol Lett 152:279–285

Takami T, Nakasone K, Hirama C, Takaki Y, Masui N, Fujii F, Nakamura Y, Inoue A (1999) An improved physical and genetic map of the genome of alkaliphilic *Bacillus halodurans* C-125. Extremophiles 3:21–28

Takami H, Nakasone K, Takaki Y, Maeno G, Sasaki R, Masui N, Fuji F, Hirama C, Nakamura Y, Ogasawara N, Kuhara S, Horikoshi K (2000) Complete genome sequence of the alkaliphilic bacterium *Bacillus halodurans* and genomic sequence comparison with *Bacillus subtilis*. Nucleic Acids Res 28:4317–4331

Tsujii K (2002) Donnan equilibria cell walls: a pH-homeostatics mechanism in alkaliphiles. Colloids Surface B Biointerfaces 24:247–252

Yamashima T (2003) Jokichi Takamine (1854–1922), the samurai chemist, and his work on adrenalin. J Med Biogr 11(2):95–102

Name Index

© Springer Japan 2016
K. Horikoshi, *Extremophiles*, DOI 10.1007/978-4-431-55408-0

Subject Index

© Springer Japan 2016
K. Horikoshi, *Extremophiles*, DOI 10.1007/978-4-431-55408-0